OUR GENETIC FUTURE

The Unintended Consequences of
Overcoming Natural Selection

William S. Blau MD, PhD

This book is dedicated to all the teachers who stimulated, educated, and encouraged me throughout my life.

CONTENTS

Title Page
Dedication
SECTION ONE: HOW WE GOT HERE — 1
Chapter 1: Origins — 3
Chapter 2: How Are Human Traits Inherited? — 10
Chapter 3: Are Humans Subject to Natural Selection? — 18
Chapter 4: Other Processes Affect Our Genome — 35
SECTION TWO: CURRENT STATUS OF THE HUMAN GENOME — 47
Chapter 5: Is Our Genome Perfect? — 48
Chapter 6: Genes, Fitness, and Disease — 63
SECTION THREE: RELAXATION OF NATURAL SELECTION — 79
Chapter 7: Human Culture and Natural Selection — 80
SECTION FOUR: IMPLICATIONS — 89
Chapter 8: Consequences of Relaxed Selection — 90
Chapter 9: Implications for Fertility — 107
Chapter 10: Implications for Infertility — 115

Chapter 11: Implications for the Immune System   128
Chapter 12: Implications for Intelligence   151
SECTION FIVE: OUR GENETIC FUTURE   163
Chapter 13: The Path Forward   164
GLOSSARY   184
Acknowledgements   209
About The Author   211

INTRODUCTION: Our Genetic Well-being: Causes for Concern

Most of my medical career was devoted to assisting patients in pain. The overwhelming majority suffered intractable low back pain, a common yet debilitating malady. There are numerous potential sources for back pain, including bones, intervertebral discs, facet joints, sacroiliac joints, nerves, muscles, tendons, and ligaments. Most physicians, however, have little to no training in distinguishing among these sources and are limited to treating the symptoms rather than the cause. A typical new patient to my practice would have seen several other physicians, often would have already undergone an operation, such as a laminectomy or a spinal fusion, and would have been prescribed a multitude of different drugs, in the most severe cases opioids. While at least 80% of us will experience acute low back pain in our lifetimes, around 8% of all adults will experience persistent or chronic pain. Low back pain is a significant cause of disability in the US, estimated to result in the loss of 149 million days of work annually. The health care costs associated with this problem are enormous – in 2016, low back and neck pain were the most expensive conditions in the US, at a national cost of approximately $134.5 billion.

This wasn't always the case. In fact, back pain only became a major health problem nationally during the latter half of the 20th century. A study from the University of North Carolina reported that the prevalence of chronic low back pain impacting function has increased in that state from 3.9% in 1992 to a whopping 10.2% by 2006. Increases are affecting adults across the spectrum regardless of age or other demographic factors. Nationwide, the escalating number of patients seeking care for back pain in recent decades has been likened to an epidemic. The causes are complex, as there are associations with low educational status, obesity, and psychosocial factors such as depression or workplace dissatisfaction. Despite the complexity, it is noteworthy that the possibility of increased genetic susceptibility for this condition has not been considered a contributing factor.

I tend to view human health issues such as this from an evolutionary perspective. After college, I studied ecology and evolutionary biology, eventually receiving a doctoral degree. My training prepared me primarily for academic work, but the university job market was quite limited. Five years after completing my doctorate, I decided that a career change was in order, and I was very fortunate to be accepted as a medical student at the University of North Carolina at Chapel Hill. After medical school, I remained at UNC while I completed a residency in Anesthesiology and Pain Medicine and subsequently joined the medical faculty until my retirement 25 years later. Although my primary occupation was providing care for each of my patients suffering from chronic pain, I was struck by the increasing number of such patients. I began to question if there was an evolutionary explanation.

With my combined expertise as an evolutionary biologist and physician, I made it my mission to discover if genetic changes might contribute to the increase in lower back pain and other chronic ailments. From an evolutionary perspective, we know that the vertebrate spine did not initially evolve to support the upright stance of a species walking on only two legs. Evolutionary alterations in the curvature of the spine have allowed us to do so, but not without the risk of excessive forces acting upon the lumbar area (low back), leading to spinal degeneration, nerve symptoms such as sciatica, and chronic pain. Imagine the impact that pain originating from the lumbar spine would have had on our day-to-day lives in the centuries before modern healthcare, analgesic drugs, spinal injections, surgical interventions, physical therapy, and psychological support. For humans in prehistoric times, strength, sturdiness, and mobility were essential attributes for survival, securing a mate, reproducing, and providing for a family. Natural selection would have acted to minimize the inheritance of any susceptibility to low back problems, especially in younger patients. So why are we seeing a nationwide epidemic of low back pain now?

Even 500 years ago, many of us would not have survived to our current age or even come close to growing old. Now, we expect to live a long and healthy life well into our 70's, 80's, and even 90's. In developed countries, medical and public health interventions are so effective at combating threats to survival that many of us live our entire lives rarely having to confront death or face even a single near-death experience. Even when our loved ones pass away, it often takes place within advanced medical facilities or hospice centers that serve as a buffer against the pain and trauma of dealing with it firsthand. But this was not the case throughout nearly all of the hundreds of thousands of years of human existence. Whether from trauma related to human conflict or simply the physical risks of a vulnerable life with numerous environmental hazards and extremes, death was a constant threat and a regular part of life as a human. Until recently, the availability of an adequate food supply or safe water was by no means guaranteed. There was also always a risk of infection that could overwhelm the immune system. Intrinsic disorders, such as diabetes - without diagnosis or medication - caused pain, suffering, disability, and death. The history of high mortality, particularly during childhood, is one of the reasons that humans had characteristically large numbers of children: to compensate for those that would likely be lost. These various threats to our lives reflected the mechanisms of natural selection favoring those best able to survive, with adaptation as a result.

Among other things, human adaptation over the millennia has led to an advanced intellect and physical dexterity that far surpasses any other species. Moreover, we have used these abilities to improve our lives in ways that circumvent the mechanisms of natural selection. The process began over 10,000 years ago but has escalated exponentially over the past 200 to 300 years. We now cultivate the food we need, construct dwellings and workplaces with tightly controlled environments, and create social structures and networks that help support the less fortunate among us. We have increased our understanding of human biology and pathology, so much so that public health measures and advanced medical care allow us to avoid or treat many of the maladies that affected us historically. Modern health care is truly a marvel of human accomplishment! We have not yet conquered every disease, but most humans today can be assured a profoundly better quality and duration of life than our progenitors. However, as this revolution in human mastery over our environment progressed, little thought was given to the fact that the world we crafted for ourselves dramatically changed the parameters of natural selection on our species. But then, what does it matter?

In this book, I will make the case that it does matter, with significant implications for the future of our genetic well-being. The proposition that the human genome (our entire collection of DNA) could deteriorate without historical natural selection is not new. This fear contributed in part to the advancement of eugenics during the early 20th century. Thankfully, many eugenic practices, ranging from forced sterilization to genocide, were ultimately recognized as ethically reprehensible. They exploited the desire to optimize our genes as an excuse for racially biased and morally repugnant interventions. Although these practices are rightfully rejected, that does not mean the genetic concerns were unfounded. Our distaste for past eugenic practices should not blind us to the potential for genetic decline, regardless of how uncomfortable it may be to acknowledge. Some of our most revered evolutionary authorities, such as Hermann Joseph Muller and Theodosius Grigorovich Dobzhansky, expressed concerns about our genetic health decades ago. Since then, profound advances in genetics shed further light on those concerns. In view of this progress, it is appropriate to reexamine whether changes in our genetic makeup in the absence of historical natural selection may be affecting our viability and susceptibility to disorders such as chronic low back pain .

In the following chapters, we will take a closer look at natural selection as it was throughout most of our history and how it has changed in recent evolutionary times. We will review some basic concepts of human genetics and explore how our DNA has been molded by natural selection as well as other processes. These considerations will lead us to some difficult questions: What are the possible long-term implications of recent changes in natural selection? Can these changes provide a partial explanation for increases in maladies such as chronic low back pain? Is the "quality" of our genetic blueprint at risk? What can/should we do to minimize the potential impact on human well-being?

Sources:

Dieleman, Joseph L. et al. 2020. US health care spending by payer and health condition, 1996-2016. Journal of the American Medical Association 23(9): 863-884.

Dobzhansky, Theodosius and Gordon Allen 1956. Does natural selection continue to operate in modern mankind? American Anthropologist 58: 591-604.

Freburger, Janet K. 2009. The rising prevalence of chronic low back pain. Archives of Internal Medicine 169(3): 251-258.

Muller, H. J. 1950. Our load of mutations. The American Journal of Human Genetics 2(2): 111-176.

Patrick, Nathan et al. 2014. Acute and chronic low back pain. Medical Clinics of North America 98: 777-789.

SECTION ONE: HOW WE GOT HERE

CHAPTER 1: Origins

To explore our genetic future, it is necessary to understand how our species came to be where we are today – both physically and genetically. Several processes that have shaped our genome are related to our history of expansion and colonization of the world. The recent revolution in DNA sequencing has provided new insights into this history, confirming, refuting, or supplementing our previous understanding based on archaeological and paleontological evidence. Early genomic data from modern humans has shed light on the origins of many contemporary populations. In fact, individuals who are interested in their heritage can now submit a DNA sample to any one of several commercial services to obtain an analysis that provides reasonably accurate information regarding their ethnic/geographic origins. More recently, we have been able to extract and enrich DNA from humans who died thousands of years ago in various locations, improving the accuracy and detail of our histories. In this chapter, I will summarize what is currently known about human origins and the colonization of the world.

We are not the only human species that has ever existed[1]. Other species, such as the Neanderthals and Denisovans, co-occurred with modern humans (Homo sapiens) as recently as 35 to 50 thousand years ago (abbreviated kya). The lineage leading to modern humans was characterized by the development of social behavior, the utilization of fire, and the development of tools that could be used as weapons. We are most likely to have diverged from our closest hominin ancestors between 260 and 350kya. The exact geographic location of our origins is not known, but it is generally believed to be within sub-Saharan Africa. Early human populations in this region were subdivided based on geography, culture, and language, with proximity and migration providing for variable levels of interbreeding. For this reason, it is overly simplistic to assume that we evolved in a linear, continuous progression over time in a single ancestral population. Instead, our features likely evolved in a mosaic of different subpopulations and lineages that were connected over time by interbreeding, thereby producing the final product that we recognize as modern humans.

For thousands of years, we were limited to the modern-day continent of Africa, where we lived as hunter-gatherers using various stone tools in a largely unmodified environment. Then, around 50-100kya, one or more relatively small groups of migrants emerged from the African continent. One of those groups, emerging around 65kya, appears to have been the primary source for the subsequent population of the entire world (Figure 2-1).

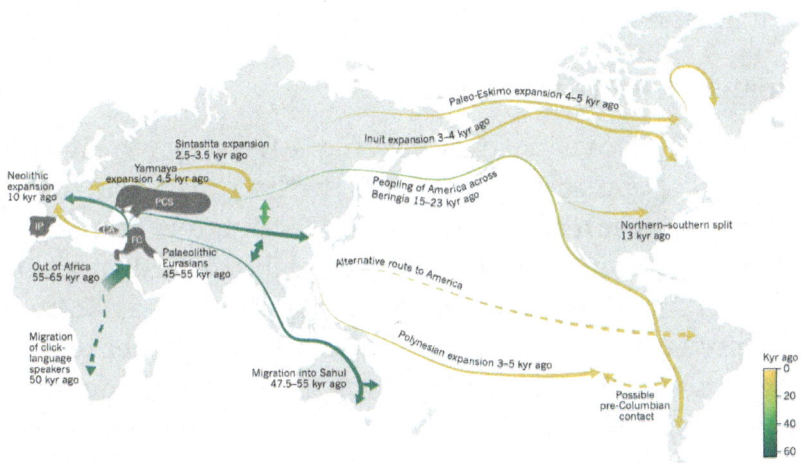

Figure 2-1. Major human migrations across the world as inferred through analyses of genomic data. Dashed lines indicate proposed routes of migration that remain controversial. CA. Central Anatolia: FC. Fertile Crescent: IP. Iberian Peninsula: PCS, Pontic-Caspian steppe. Reproduced with permission from Nielsen, Rasmus 2017. Tracing the peopling of the world through genomics. Nature 541: 306. Springer Nature.

Early populations outside of Africa continued to follow a hunter-gatherer lifestyle until the agricultural transition around 11kya, when a new way of life began to emerge in several parts of the world, particularly in the Fertile Crescent region of Southwest Asia. This new way of life involved a more sedentary existence, with agriculture and animal husbandry replacing hunting and gathering as the primary means of human sustenance. The stability afforded by the domestication of plants and animals led to the development of more permanent settlements supporting larger populations. At the same time, the crowding of people in conditions that were often unhygienic with poor sanitation led to widespread infectious diseases. In addition, unreliable supplies of carbohydrate-rich, poor-quality foods increased the risk of nutritional deficiencies.

Expansion And Colonization

About 23,000 years after leaving Africa, some of the founding population expanded to the north and west while others spread to the east and ultimately colonized Australasia and New Guinea. Although the first modern humans may have arrived in Europe as early as 45kya, this predated an advance of glaciers which peaked at 25kya, and those populations appear to have made little genetic contribution to present-day Europeans. As the climate improved, the region was recolonized by hunter-gatherers. Farming in Europe began about 8kya with the influx of populations from the Near East, where the agricultural transition had already occurred. Through mass migration and assimilation of local hunter-gatherers, the transition reached as far as Britain and Scandinavia by 6kya. This was followed by a massive migration of herders from the Eurasian Steppe at about 4.5kya, extending to western and northern Europe. Consequently, most present-day Europeans are derived from West European hunter-gatherers, ancient North Eurasian pastoralists, and early European farmers of Near Eastern origin.

    The ancestors of Native Americans arrived from far northeast Asia in at least three waves via a land bridge known as Beringia that connected present-day Russia and Alaska. Two of these waves were limited to the Arctic regions. The third is believed to have been derived from a single founding population that split from their Siberian ancestors at about 35kya. Genetic evidence suggests that the branch leading to all subsequent North and South Americans was derived from only five founding maternal lines, which were, in turn, derived from a common ancestor at about 15kya to 18kya. A subsequent split about 13kya led to northern and southern branches. Movement continued south along the west coast of North America, likely followed by northerly back-migrations through the interior.

Eastward expansion out of the Near East may have occurred in one or two early waves, reaching Oceania around at least 47 to 55kya. Polynesian expansion continued until about 3 to 5kya, possibly reaching the Americas and interbreeding with Native Americans.

The way in which the range of human habitation expanded and new territories were colonized had profound effects on the present-day genetic structure of human populations. We know that contemporary inhabitants in many locations often do not reflect, in any simple way, the ancient inhabitants of the same area. Migrations occurred over long distances. In some cases, migrants encountered pre-existing human populations from prior colonizations. If the new migrants succeeded in expanding into this new territory, they may have replaced the existing populations, interbred with them, or done both to some degree. In fact, nearly all contemporary populations demonstrate genetic evidence for past interbreeding with other groups.

Bottlenecks occur when a population goes through a period of markedly reduced size and were a common occurrence during range expansions. For example, it is believed that the original group that emerged from Africa represented perhaps one-fifth the size of the African source population of nearly 14,000 individuals. This means that they carried a relatively small and likely random selection of the total genetic variability contained in the source population. Evidence shows that a second pronounced bottleneck occurred around the time of the split of European and Asian populations. Other bottlenecks occurred as humans moved into new, unoccupied territories, for example, when the Siberian ancestors of Native Americans migrated into the Americas. The serial founder model suggests that human expansion generally involved numerous bottlenecks as small bands of individuals repeatedly pushed into unoccupied territories at the expanding front of human occupation. Bottlenecks and serial founder effects can have a significant impact on the genetic composition of subsequent generations, as we will see in Chapter 4.

Sources:

Bergström, Anders 2021. Origins of modern human ancestry. Nature 590: 229-237.

Nielsen, Rasmus 2017. Tracing the peopling of the world through genomics. Nature 541: 302-310.

Scerri, Eleanor M. L. 2018. Did our species evolve in subdivided populations across Africa, and why does it matter? Trends in Ecology and Evolution 33(8): 582-594.

Wolfe, Nathan D. et al. 2007. Origins of major human infectious diseases. Nature 447: 279-283.

CHAPTER 2: How Are Human Traits Inherited?

Our goal is to examine the nature of natural selection in modern human populations and understand its role in shaping past adaptations and current challenges. For adaptation to occur, there must be variability among individuals in traits that affect fitness[2], and at least a portion of that variability must be heritable[3]. Since heritability is dependent upon genes, it is genetic variation that provides the raw material for adaptation and, ultimately, evolution. Understanding the basic structure and function of our genes is helpful to fully appreciate the impact of modern culture, technology, and healthcare on our species. In this chapter, I will present an overview of basic genetics to serve as a reference for subsequent discussions.

DNA, Genes, And Chromosomes

When you observe a human, what do you see? How tall are they? What color is their hair? Are they overweight? A doctor may ask, what is their blood pressure? What is their blood sugar? What is their percentage of body fat? These observable or measurable traits are called the person's phenotype, which results from interactions between the person's genes and their environment. It contrasts with the genotype: an individual's specific, heritable genetic makeup that contributes to their phenotype. But what are genes, exactly? A gene is a specific bit of DNA (deoxyribonucleic acid) that contains instructions for a particular phenotypic trait, including all of the structures and compounds necessary for sustaining life. Most of your genes are in chromosomes[4] found in nearly all the body's cells. Humans have 23 pairs of chromosomes, one of each pair being derived from the individual's mother and one from the father. The two chromosomes that make up a pair contain the same array of genes (except for the sex chromosome pair X and Y), so we carry two copies of each gene.

Unraveling the details of DNA structure and function has been one of the most exciting developments in biology over the past century. It was isolated for the first time in 1868, but it was the early 1950s before it became generally accepted that DNA was the genetic material accounting for the inheritance of traits. In their landmark publication in 1953, James Watson and Francis Crick from the University of Cambridge described the structure of DNA[5].

DNA exists in very long strands. If the DNA in a human cell were stretched out, it would be about 6 feet long, and all the DNA in the human body would stretch to the sun and back about four times! The backbone of each strand is a chain of sugar molecules (deoxyribose) joined together by phosphates. Attached to each sugar molecule is one of four possible bases: adenine, guanine, cytosine, or thymine – usually designated as A, G, C, or T, respectively. These bases project out from the sugar-phosphate axis of the chain. The combination of the sugar+base+phosphate is called a nucleotide. Imagine the long chain of nucleotides, each with a base extending off to the side. These bases can pair up with complementary bases (A pairs up with T, and G pairs up with C) from a second nucleotide chain, creating a ladder-like structure with two sugar-phosphate "rails" on the outside and attached pairs of complementary bases on the inside forming the "steps" of the ladder. Now imagine this whole structure twisted into a helix forming something like a spiral staircase, and you can visualize the structure of DNA.

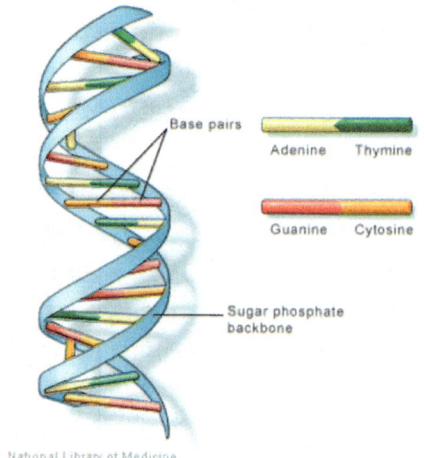

Figure 3-1. Structure of DNA. US National Library of Medicine

These helices, in turn, are further coiled in a compact fashion in the chromosome. Genes are functional segments of DNA, each one containing, on average, a sequence of about 3,000 nucleotides (each one part of a base pair). Each DNA molecule has hundreds of thousands of nucleotides, and the entire human genome contains over 3 billion nucleotides.

How Do Genes Contribute To Phenotypes?

Many genes act by directing the production of proteins. Proteins are complex and often very large molecules that serve many essential roles in the body. Some regulate basic chemical processes within the cells, while others have a structural purpose, including prominent physical features such as hair. Proteins are made up of long chains of hundreds to thousands of smaller molecules called amino acids. There are 20 different amino acids, leading to vast numbers of different possible combinations that allow for tremendous diversity in protein structure. Depending upon the amino acid sequence, the chains that compose protein molecules will fold into complex shapes and connect in different ways that influence how the protein works.

The DNA content of genes directs the assembly of proteins by determining the sequence of the amino acid building blocks. The sequence of bases within the DNA strand is actually a code – the genetic code. Every group of three bases constitutes a triplet that directs the addition of a specific amino acid into the protein chain. For example, the sequence ATG codes for the amino acid methionine. There is some redundancy, however, as ATT, ATC, and ATA all code for the amino acid isoleucine[6].

The first step in building a protein requires creating a copy of the gene containing the genetic code for that protein. However, this copy differs in one way; the sugar in the chain of this molecule is ribose rather than deoxyribose. For this reason, the new molecule is called ribonucleic acid, abbreviated RNA. RNA contains the entire base sequence of a gene, but not all the bases contribute to the genetic code that determines the protein's amino acid sequence. The segments that do not contribute are removed by a chemical editing process so that only the coding segments are left by the time the protein is synthesized. The RNA is transported to cell structures called ribosomes, which "translate" the base code and carry out the instructions as they add amino acids to the developing protein molecule, thereby linking the genotype to the phenotype.

Other Characteristics Of Human Genes

Humans possess about 22,000 protein-coding genes, accounting for less than 2% of our genome. The exact function of the remaining DNA has long been debated. It has even been called "junk DNA," stemming from the belief that it has no function. However, we now know that some genes within this non-coding portion of our DNA have important functions, including regulating the activity of other coding genes. In fact, it is estimated that evolutionary adaptations are over 10-fold more likely to involve alterations in gene regulation rather than protein structure. Most regulatory genes in a population have DNA variants (see below, some of which can have greater than two-fold effects on the amount of protein produced by the coding genes they regulate.

There are other features of genes that are relevant for understanding the effects of natural selection. As we have seen, each of our 23 pairs of chromosomes contains one chromosome derived from our mother and one from our father. These chromosome pairs (excluding the X and Y sex chromosomes) contain the same array of genes; however, they are not always identical. They may code for the same protein but with a slightly different amino acid structure due to variations in the genetic code (i.e., base sequence) they contain. Such alternate versions of a gene are called alleles. The ABO gene that determines the phenotype 'human blood type' is a good example. The maternal and paternal contributions to this gene will each contain one of three possible alleles: A, B, or O. If a person ends up with two copies of the same allele, such as A+A, they are homozygous for that gene variant, and their blood type will be A. If they carry different alleles, such as A+B, they are heterozygous, and their blood type will be AB. In some cases, the presence of one allele will mask the presence of another. For example, a heterozygote[7] carrying A+O will have blood type A. In this case, the A allele is said to be dominant, while the O allele is recessive. The B allele is also dominant to O, so the only way a person can have the O blood type is if they are homozygous with two copies of O.

ABO genotype in the offspring		ABO alleles inherited from the mother		
		A	B	O
ABO alleles inherited from the father	A	A	AB	A
	B	AB	B	B
	O	A	B	O

Table 3-1. Summary of the inheritance of the ABO blood types. US National Library of Medicine.

There is one other characteristic of genes that can have significant effects on selection and evolution. It is quite common for one gene to be responsible for more than one trait. For example, Marfan syndrome is a medical disorder that results from an abnormality in a single gene and is characterized by several traits that are not clearly associated with one another: increased height, thin fingers and toes, dislocation of the lens of the eye, and increased risk of aortic aneurysm. A gene that influences multiple traits is said to be pleiotropic. One study found that 44% of genes are pleiotropic, with the number of phenotypes per gene ranging from 1 to 53.

Sources:

Chesmore, Kevin, et al. 2018. The ubiquity of pleiotropy in human disease. Human Genetics 137: 39-44.

CHAPTER 3: Are Humans Subject To Natural Selection?

"...natural selection has a bright and a dark face. The bright face reflects all the beauties, and exquisite adaptations, of the living world. The dark face hides the selective culling of the less fit in parasite-ridden lives that are often nasty, brutish, and short."[8]

A Closer Look At Natural Selection

In the history of our species, natural selection has had a profound influence on shaping and maintaining the human genome. Most people are generally familiar with the term natural selection, but 'survival of the fittest' provides minimal insight into its nature and importance. Regrettably, even teaching this subject in our schools has become controversial, so I will review some basic concepts that pertain to our subsequent discussion.

Charles Darwin's concept of natural selection was largely influenced by his observations on artificial selection: the selective breeding performed by humans on other species. Plants or animals with traits thought to be desirable – say, vegetables with increased cold-hardiness – are selected for fertilization, with the result that offspring are more likely to also be cold-hardy. If selective breeding is consistently performed over generations, cold-hardy plants will come to predominate. The clearest example of this is the incredible variety of dogs humans have bred over numerous generations to perpetuate differences in personalities, physical appearances, and skills such as hunting or herding.

Natural selection differs in that the favored traits are not chosen arbitrarily; instead, they are those characteristics that enhance survival and reproductive success relative to others in a population. This is a passive process, with no specific goal imposed on the outcome. In any system that replicates and has variability, individual lines that produce viable "offspring" faster and/or in greater numbers than others will come to predominate. Imagine the lineage of a hypothetical mouse species. We will start with a female drawn from a population where litters always include two females. Our female breeds (we will assume that there are always adequate numbers of males for this purpose), but one of her two female progeny carries a new genetic mutation that allows her to produce four viable female offspring (a high breeder), whereas the other will produce the usual two (a low breeder). Let us also assume that the genetic tendency to produce four vs. two female offspring is always passed along to female progeny. That means that in the next generation, the high-breeding female will contribute four female mice to the total population while the other will contribute only two. By the time we reach ten generations, the high-breeding genetic line has generated 1,048,576 female mice, whereas the low-breeding line has generated only 1,024 female mice. In the first generation, with two females, the proportion of high breeders to low breeders is 1:1. After ten generations, it is 1024:1. The genetic composition of this mouse population has changed and now reflects mostly high breeders. The mutation that enabled the production of four female offspring increased the fitness of the individuals carrying it. Consequently, those individuals came to predominate through the process of natural selection. Not only that, but evolution has also occurred so that the average member of this population now produces nearly four offspring, rather than the two of the original population. This scenario is highly oversimplified relative to real-life biology, but it demonstrates natural selection's powerful yet passive nature. There was no design to develop high-breeding mice. Nobody chose the high-breeder mice and their genes to predominate. They "chose" themselves due to their increased ability to reproduce. You would find the same kind of result if genetic changes led to an increase in survival to reproductive age or earlier age at reproduction rather than an increase in the number of offspring.

We can now define natural selection as the process by which a heritable trait becomes either more or less common in a population based on the relative survival and reproductive success (= ecological "fitness") of those with the trait compared to others of the same species. In the simplest case, a genetic variant associated with greater fitness becomes "fixed" and possessed by all, while other variants eventually disappear from the population. This leads to adaptation - the evolutionary process whereby members of a population become better suited to survive and reproduce in their environment.

An interesting example of recent selection involves the African savannah elephant populations in

Mozambique's Gorongosa National Park. Heavy poaching of elephants for their ivory tusks during a civil war in this area between 1977-1992 resulted in a decline of around 90% in the elephant population. During the same period, there were significant changes among the remaining elephants. The frequency of tusklessness (absence of tusks) increased markedly, from about 18.5% before the civil war to 50.9% afterward, but only in females. In addition, there were very few adult males compared with the number of females. Among the offspring of the surviving females, tusklessness remained elevated at 33%. Recently, it has been determined that a dominant mutation on the X chromosome is a likely explanation for these findings. Females who carry this mutation on at least one of their X chromosomes will be tuskless, and in the face of intense poaching, such females are estimated to have survived at a rate around five times higher than that for females who had tusks. Apparently, selection favoring the "tuskless" mutation resulted in females with that mutation predominating in the population since they would be useless to poachers seeking tusks. It turns out that the same mutation in males results in early death, accounting for their relative scarcity. It can be argued that this is a case of artificial selection since the driving force was human poaching. However, the poachers were not seeking to promote tuskless females. But they did alter the elephants' environment such that elephants with tusks experienced greater mortality, resulting in a selection pressure favoring the tuskless condition.

    To what extent have natural selection and adaptation affected our human species? Historically, the causes of our morbidity and mortality - infection, disease, trauma, starvation, and exposure – were agents of natural selection in the pre-technological world. With the exponential growth of genetic data that has become available since the first publication of the human genome in 2001, we now have ample evidence for a history of adaptive human evolution in response to natural selection affecting over 4,000 genes involved in various processes. There is also consistent evidence for relatively recent selection affecting genes involved in adaptation to diverse physical environments, pathogens, and changing dietary practices. Let's look at some examples.

Adaptation To The Environment

The humans who emigrated from Africa must have faced numerous environmental challenges as they colonized the globe and encountered conditions that sometimes differed dramatically from those under which they initially evolved. For example, natural selection should have favored adaptation to colder temperatures as populations migrated north. A multi-institutional study published in 2018 explored this possibility by analyzing a gene designated TRPM8 in Eurasian populations at different latitudes. This is the only gene known to be associated with the ability to sense moderate cold temperatures and the subsequent physiological response necessary to maintain core body temperature. A second genetic site is believed to play a role in regulating the activity of TRPM8, and one variant at this site exhibited evidence of positive selection during the last 25,000 years. Positive selection occurs when one genetic variant confers greater fitness and comes to predominate or replace alternative variants in a population over time. The strength of this selection correlates with latitude;

its frequency in populations varies from 5% in Nigeria to 88% in Finland. The authors of this study believe that the favored variant arose in Africa, where it would have been neutral with regard to natural selection, and subsequently underwent increasingly strong positive selection as populations colonized further northward[9].

Another remarkable example of human adaptation to the environment involves the indigenous Bajau people of Southeast Asia, known as 'Sea Nomads.' For over 1,000 years, their marine hunter-gatherer lifestyle has depended upon the ability to spend most of their working time holding their breath for long periods of time and diving to collect food. This ability is dependent upon a physiological "diving response," shared by humans and other organisms, that serves to reduce oxygen consumption and prioritize blood flow to the organs that are most susceptible to the adverse effects of oxygen deprivation. As part of this response, oxygenated red blood cells that are sequestered in the spleen can also be released into the general blood circulation to enhance oxygen delivery to the critical organs.

An analysis of the genome of the Bajau people has revealed evidence for selection affecting several genes, at least two of which appear likely to be related to adaptations to the oxygen stress associated with breath-hold diving. One of the gene variants that has undergone strong selection is thought to enhance the part of the diving response that allocates blood flow to the most oxygen-dependent organs like the brain, heart, and lungs. A selected variant of the other gene is associated with larger spleen size, with the potential to hold more red blood cells in reserve. The same variant also affects the contraction of the spleen – the process involved in releasing those blood cells into general circulation. It seems likely that individuals who possess these mutations are more successful at breath-hold diving and food collection, resulting in relatively improved survival and reproduction. Persistent selection for these attributes over time has resulted in the ability of the Bajau people to excel at their particular lifestyle.

One additional example involves the first evidence of human adaptation to an environmental toxin. Indigenous populations of the Andean highlands have consumed water naturally contaminated with arsenic for thousands of years, even though arsenic is a well-known toxin causing nearly every major organ system to dysfunction. A Scottish study published in 2015 found that inhabitants of the northern Argentinian Andes have a unique mechanism of arsenic metabolism that is more efficient and faster, limiting exposure to toxic metabolites. A genetic analysis[10] found a strong association between a specific gene designated AS3MT and urinary arsenic metabolite levels, suggesting that this gene plays a key role in arsenic metabolism. The genetic region near AS3MT reveals evidence for past selection, and it is postulated that genetic variants with a regulatory role in AS3MT function may have been the target. The analysis also found that the specific genetic variants around AS3MT were different from those in other related populations where arsenic levels were not as high. The authors conclude that high levels of arsenic have imposed selective pressure on local Andean populations, leading to an increase in protective genetic variants associated with the gene involved in arsenic metabolism, altering metabolism to minimize toxicity.

Adaptation To Pathogens

Infectious diseases have imposed profound selective pressures throughout human history, particularly since the transition from hunter-gathering lifestyles to farming about 10,000 years ago. Examples of specific genetic adaptation have been documented for resistance to prions (abnormal pathogenic proteins), viruses, protozoa such as *Plasmodium vivax* (malaria) and *Trypanosoma brucei rhodesiense* (African sleeping sickness), bacteria such as *Vibrio cholerae* (cholera), and larger parasites. Malaria has been, and still is, a significant agent of selection in many human populations. The genetically based resistance to malaria provides one of the best understood and most interesting examples of human adaptation.

One significant mechanism of resistance to malaria involves the gene responsible for the production of hemoglobin – a protein carried by all red blood cells. Hemoglobin is essential for red blood cells to pick up oxygen from the lungs and distribute it throughout the body. One particular mutation in the hemoglobin gene leads to an alteration in the protein, which renders the individual carrying it to be resistant to the parasite that causes malaria. This genetic variant is known to have been subject to a form of positive selection but has never come to predominate in our genome. Why should that be?

Recall that we all carry two versions of nearly every gene – one derived from our mother and one from our father. If the two versions of a gene carried by an individual are identical, that person is a homozygote for that gene[11]. If a person is homozygous for the normal hemoglobin gene, they will have no problems with oxygen transport but will be susceptible to malaria. If a person is homozygous for the malaria-resistant variant, all the hemoglobin is abnormal, impairing oxygen delivery and causing red blood cells to assume a sickle shape. These cells impair blood flow – basically clogging up small blood vessels – leading to the syndrome of pain and organ damage known as sickle cell disease. Over time, you would expect the abnormal hemoglobin variant to disappear from the population due to selection against it. But this is not what occurs. People who carry one copy of the normal gene and one copy of the variant (heterozygote, or carrier for sickle cell disease) will have enough normal hemoglobin to avoid the hallmarks of sickle cell disease but benefit from resistance to malaria. In regions of the world where malaria is prevalent, this is a considerable advantage. In evolutionary terms, the heterozygote is more fit than either homozygote. However, each heterozygote will be equally likely to pass along either the normal gene or variant gene to its offspring, so both types of genes persist in the population - even though the unfortunate individual who ends up homozygous for the abnormal hemoglobin variant will have sickle cell disease.

There is more to the malaria story. Kimberly McManus of Stanford University and her colleagues have reported that the ability of the protozoan responsible for malaria to infect host cells depends upon the presence of a particular protein on the surface of those cells. The protein is coded for by a gene that has been designated DARC (Duffy antigen receptor for chemokines). There are three classes of genetic variants for this gene, and one of them protects against infection, presumably by altering the host protein. This variant is found in nearly 100% of individuals from sub-Saharan Africa, where malaria is common, whereas the other two variants are common elsewhere. There is evidence that the malaria-resistant variant arose 42,000 years ago and was subject to positive selection pressures among the strongest known within the human genome.

Adaptation To Diet: Got Milk?

One of the best documented examples of positive selection in humans involves our ability to digest lactose, a sugar in milk that serves as an essential energy source for infants. The production of the enzyme lactase in newborns' small intestines is essential for breaking down and digesting lactose. In our evolutionary past, the production of lactase ceased after the age of weaning. Consequently, adults were unable to break down or absorb lactose and had no need to since the adult diet of our hunter-gatherer progenitors did not include milk. All of this changed with the domestication of dairy animals, which is thought to have first occurred in a region of present-day Turkey about 10,500 years ago. As farmers migrated, the practice spread to Europe beginning around 9,000 years ago and into Africa around 7,000 years ago.

If adults lacking an active lactase gene consume milk, they will experience typical symptoms of lactose intolerance. When lactose proceeds through the gut undigested, the bacteria inhabiting our colon will exploit it, resulting in symptoms of abdominal discomfort, gas, bloating, and diarrhea. In about one-third of all humans today, lactase production persists so that adults can consume milk and milk products without intolerance symptoms. However, there is wide variation in the frequency of this trait among different human populations, ranging from about 5% to almost 100%. The frequency has been found to be highest in people of northern European descent and some populations from West Africa, East Africa, India, and the Middle East - populations that historically incorporated large amounts of milk into their diets after adopting pastoral lifestyles herding cattle, sheep, goats, and other livestock[12].

Genetic evidence indicates that selection favoring lactase persistence into adulthood may have been among the strongest known for human traits over the past 10,000 years. The precise nature of the selective advantage of adult milk consumption has yet to be known for certain but may be related to milk providing an uncontaminated, high-quality, and reliable source of food and fluid. It is also possible that the increased levels of calcium and vitamin D that milk contains were beneficial in higher latitudes where sunlight, essential for vitamin D synthesis, was limited. A more recent multi-institutional study published in 2022 proposed that the selective advantage of lactase persistence may have been episodic in times of famine or increased exposure to pathogens. What we now consider minor health effects of lactose intolerance, especially diarrhea, could lead to fatality in those times.

The genetics of lactase persistence is relatively well understood for some populations. The first known mutation was found on chromosome 2. It is not part of the lactase gene but appears to work by increasing the gene's activity, leading to greater lactase production. The frequency of this genetic variant ranges from 6% to 36% in Eastern and Southern Europe, from 56% to 67%

in Central and Western Europe, and up to 73% to 95% in the British Isles and Scandinavia. Estimates for the date of its origin include the period when dairy herding first emerged. Beyond Europe, it has also been found to account for lactase persistence in parts of Asia and Central and North Africa, but interestingly, the genetic basis for lactase persistence seems to be different in East Africa. There, three alternate mutations have been identified that also act by affecting the lactase gene's activity. These genetic variants date to about 7,000 years ago and provide a remarkable example of convergent evolution – when independent evolutionary pathways lead to similar adaptations.

Different Kinds Of Natural Selection

Most of the examples discussed in the previous section demonstrate positive selection. This is the type of situation that most of us are likely to envision when we think of natural selection. However, other forms of selection may be equally or even more important under certain circumstances.

Balancing selection is another form of natural selection that serves to maintain genetic variability. We will look at one type of balancing selection called heterozygote advantage. Imagine a gene where there are two variants: a and b. Each one will have its own effect on fitness, and one may therefore be favored by natural selection over the other. But chromosomes come in pairs. What happens if one chromosome carries the a variant, and the other carries the b variant, i.e., the heterozygote ab? The heterozygote may be associated with fitness that is greater than, lesser than, or equivalent to either of the homozygotes aa or bb. In cases where the heterozygote ab results in superior fitness, it will be favored by natural selection. Since progeny of the heterozygous individual will receive alleles a or b with equal probability, both variants will be maintained in the population. This is precisely what happens with the hemoglobin gene associated with sickle-cell disease in the example discussed earlier. Balancing selection is believed to play a significant role in maintaining variability in many of the genes underlying our immune system.

A significant misconception about natural selection is that it primarily results in change and adaptation. In fact, one of the most profound ongoing effects of natural selection is to maintain the "status quo." New mutations impairing survival and/or reproduction will be at a selective disadvantage and tend to be eliminated from the population. This is called *negative, or purifying, selection.* Harmful mutations will not accumulate and cause degradation of the genome because those individuals who carry them will be less fit and, therefore less able to contribute their mutations to subsequent generations. Natural selection has shaped and optimized our genetic makeup, acting upon thousands of past generations. It tends to preserve an optimal genotype and minimize the frequency of genetic variants, or combinations of variants, that depart from this optimum. In this way, genes that underlie many crucial physiological processes are found to be genetically "conserved," exhibiting little or no variation. In humans, it has been estimated that 75% of all mutations that have resulted in an alteration of amino acids (the building blocks of proteins) have been eliminated by purifying selection. In general, the relatively few new variants that have large effects are more likely to be harmful, so their frequencies are the most severely restricted.

The history of tuberculosis in Europe provides a striking example of negative selection in humans. Tuberculosis (infection by the bacteria *Mycobacterium tuberculosis*) has caused more than one billion deaths worldwide over the past 2,000 years and continues to be a significant public health threat today. It has been determined that a recessive variant of a gene designated TYK2 (for tyrosine kinase 2) can disrupt mycobacterial-specific immunity, rendering individuals who are homozygous for this variant more susceptible to tuberculosis. The authors of a French study published in 2021 investigated the evolutionary history of this variant in Europe over the past 10,000 years by analyzing over 1,000 genomes derived from ancient human remains. Based on this analysis, the variant appears to have arisen as a single mutation about 8,500 years ago. Over time, the frequency of this variant in Europe gradually increased, likely in a random fashion, to a maximum of about 10% around 3,000 years ago. However, beginning about 2,000 years ago, the frequency was found to consistently decrease, currently averaging at 2.9% in Europe today. The authors attribute this to strong negative selection against the variant in the face of increasing prevalence of clinical infection with tuberculosis, with an estimated 2.5 million deaths over the past 2,000 years in individuals that were homozygous for this recessive variant. Homozygous individuals are believed to experience an estimated 20% reduction in fitness compared to individuals with only one or no copies of this variant.

Thus far, we have focused on simple genetic traits for the purposes of illustration. These and others, such as colorblindness, hairline shape (presence or absence of a widow's peak), and dimples, seem to be determined by variants of a single gene. However, many human traits, such as height or intelligence, are determined by multiple variants across numerous genes, hundreds or thousands in some cases, each with a relatively small individual effect. There is abundant evidence for the action of natural selection on such traits, although its effect on each individual variant is less profound and varies with the influence of each variant on overall fitness. Traits that are beneficial and influenced by variants among multiple genes may undergo polygenic adaptation, the simultaneous positive selection of many genes. This may be the most common type of selection that occurred as human colonization of the world subjected us to new environments and selective pressures; for example, evidence indicates that adaptation to high altitude has a polygenic basis.

Sources:

Campbell-Staton et al. 2021. Ivory poaching and the rapid evolution of tusklessness in African elephants. Science 374: 483–487.

Darwin, Charles 1859. On the Origin of Species. John Murray: London.

Evershed, Richard P. et al. 2022. Dairying, diseases and the evolution of lactase persistence in Europe. Nature 608: 336-345.

Field, Yair et al. 2016. Detection of human adaptation during the past 2000 years. Science 354 (6313): 760-764.

Fu, Wenqing and Joshua M. Akey 2013. Selection and adaptation in the human genome. Annual Review of Genomics and Human Genetics 14: 467–489.

Guo et al. 2018. Leveraging GWAS for complex traits to detect signatures of natural selection in humans. Current Opinion in Genetics & Development 53: 9–14.

Ilardo M. A. et al. 2018. Physiological and genetic adaptations to diving in sea nomads. Cell 173(3): 569-580.e15. https://doi.org/10.1016/j.cell.2018.03.054.

Kerner, Gaspard et al. 2021. Human ancient DNA analyses reveal the high burden of tuberculosis in Europeans over the last 2,000 years. The American Journal of Human Genetics 108: 517-524.

Key, Felix M. 2014. Advantageous diversity maintained by balancing selection in humans. Current Opinion in Genetics & Development 29: 45–51.

Key, Felix M. 2018. Human local adaptation of the TRPM8 cold receptor along a latitudinal cline. PLOS Genetics 14(5): e1007298. https://doi.org/10.1371/journal.pgen.1007298.

Li, Mulin Jun et al. 2013. dbPSHP: a database of recent positive selection across human populations. Nucleic Acids Research 42: D910–D916.

McManus K. F. et al. 2017. Population genetic analysis of the DARC locus (Duffy) reveals adaptation from standing variation associated with malaria resistance in humans. PLOS Genetics 13(3): 1–27.

Rees, Jasmin S. et al. 2020. The genomics of human local adaptation. Trends in Genetics 36: 415-428.

Schlebusch, Carina M. 2015. Human adaptation to arsenic-rich environments. Molecular Biology and Evolution 32(6): 1544-1555.

Segurel, Laure and Celine Bon 2017. On the evolution of lactase persistence in humans. Annual Review of Genomics and Human Genetics 18: 297–319.

Zeng, Jian et al. 2018. Signatures of negative selection in the genetic architecture of human complex traits. Nature Genetics 50: 746–753.

CHAPTER 4: Other Processes Affect Our Genome

It is important to understand the current status of the human genome and how it came to be, to better anticipate the effect of changes in natural selection on our genetic future. Although natural selection is important, it is not the only process that has influenced our genome. Several other factors derive from our demographic history.

Population Size

One of the most significant population characteristics is size. Population size profoundly affects the likelihood of random changes in gene frequencies unrelated to natural selection. To illustrate this effect, consider a simple flip of a coin performed thousands of times. Assuming the coin is appropriately weighted to allow for an equal probability of landing on either side, the ratio of heads:tails after thousands of trials should be very close to 1:1. Any random departures from this ratio tend to cancel out over time. However, when the number of trials is reduced to, say, twelve, there is a much greater possibility that the observed ratio of heads:tails may be 2:1, 3:1, 1:2, or 1:3, just by random chance. Random effects can also influence the transmission of genetic information from one generation to the next.

For example, consider a hypothetical gene for hair color that has two alternate variants: one of them codes for red hair, and the other codes for blond hair[13]. Suppose natural selection does not favor one hair color over another. In that case, a parent carrying both variants (i.e., a heterozygote for that gene who may have, say, brown hair) will have an equal chance of transmitting either the red or blond hair variant to the next generation. In a hypothetical large population of heterozygotes with many breeding couples, the relative proportion of red and blond hair variants transmitted to the next generation will be very close to 1:1. But in some cases, especially in small populations with fewer breeding couples, the proportion of the variants transmitted may be significantly different from 1:1 just due to random chance. It is even possible for random effects to allow a single variant to predominate and eliminate all other alleles at the same gene so that everyone is a redhead.

Some of these variants may be neutral with regard to fitness and can vary randomly without consequence, but even harmful variants can potentially become more common in this manner if the population size is small enough. Random genetic drift refers to changes in the frequencies of genetic variants over generations largely due to chance.

Ultimately, the frequency of any particular genetic variant in a population is mainly determined by the balance between the input of new mutations, the effects of natural selection, and the impact of random genetic drift. As we have just seen, random effects tend to be more significant in smaller populations. But it is also true that natural selection is less effective in these populations so that random effects tend to predominate. Population genetics theory predicts that smaller populations will end up with less diversity due to the random loss of variants; at the same time, they will accumulate a greater number of harmful variants over time due to the diminished effect of natural selection. As it turns out, small populations occurred frequently throughout human history.

Bottlenecks And Founder Effects

Bottlenecks occur when a population experiences a marked reduction in size (perhaps because of disease outbreaks, famines, floods, human conflict, or expansion into new territories), which may be followed by renewed growth, in that case, resembling more the shape of an hourglass. In human history, a very significant bottleneck occurred when our species migrated out of Africa at about 65kya. Due to the relatively small size of this migrating group, it was unlikely to reflect the entire genetic pool of the source populations on the African continent. Consequently, random genetic drift and new mutations that occurred along the way significantly affected the genetic characteristics of subsequent generations that ultimately colonized the world. The impact of this event remains evident today: there is much greater genetic diversity among Africans than among members of populations from other parts of the world. In fact, a pair of chromosomes from someone with direct African ancestry will be more different from one another than two chromosomes from any two people with ancestry from elsewhere in the world.

The case of Ashkenazi Jews (AJ) provides a relatively well-studied example of the effect of bottlenecks. AJ represent a distinct branch of Judaism whose ancestors were derived mainly from populations in present-day Germany, with genetic evidence for origins in the Middle East. Historically, they would have been impacted by the bottleneck associated with migration out of Africa and at least three other bottlenecks. The first occurred at the beginning of the Jewish Diaspora around 70 CE when small numbers of individuals derived from a well-established Jewish population in the Middle East made their way to Western Europe. Populations remained there and may have grown to around 100,000 or more individuals. A second bottleneck was associated with the colonization of northern Europe during the medieval period. A severe third bottleneck occurred between 1400 and 1100 CE and was associated with persecutions during the Crusades and mortality related to the Black Death. Subsequently, renewed population growth occurred up until the 20th century[14]. Present-day AJ populations are known to harbor genetic variants for recessive diseases such as Tay-Sachs disease and Gaucher disease. In addition, there are high frequencies of genetic variants associated with increased risk for common disorders such as Parkinson's, breast, and ovarian cancer. When compared with contemporary European populations, Ashkenazi Jews have 47% more novel variants (i.e., those that are completely absent from the European population) per genome, with a slightly higher overall load of mutations. These findings are consistent with the genetic effects anticipated from historical bottlenecks.

The migration out of Africa and the Jewish Diaspora are examples of a particular type of bottleneck called a founder effect, where the first few colonizers of a region determine the subset of genes to be passed down within the local population. Because the founding population size is typically small, the effects of random genetic drift tend to be more prominent than natural selection, so even harmful variants may increase in frequency by random chance. Many other instances of founder effects have been documented. For example, due to a founder effect associated with the colonization of Finland, Finnish populations contain more rare and low-frequency harmful alleles and a higher proportion of variants associated with a complete loss of protein function compared with non-Finnish Europeans.

Human colonization of the globe is believed to have involved repeated or serial founder effects, as small populations continued to colonize new regions during the geographic expansion following migration out of Africa. The French-Canadian population provides a well-documented example of serial founder effects. Most of the current population, numbering over eight million, is derived from about 8500 French settlers who initially established colonies along the Saint Lawrence River between 1608 and 1759. Only 2600 of these settlers present before 1680 account for about two-thirds of the genetic makeup of the current population[15].

The English gained control of this territory in 1759, ending the influx of settlers from France and rendering the already-established communities isolated due to linguistic, cultural, and religious differences. Nevertheless, the French-Canadian population experienced significant growth, expanding by over 700% in less than 20 generations. Along with growth came repeated expansion into new territories within the Saint Lawrence Valley, producing founder effects followed by localized population growth in the newly colonized areas. In one instance, around 600 settlers from the original founding population around Quebec entered a new region to the northeast along the St. Lawrence River called Charlevoix between 1675 and 1850 (#1 in Figure 5-1). This and subsequent founder effects accrued with migrations from Charlevoix to the Saguenay Valley and ultimately to Lac St.-Jean.

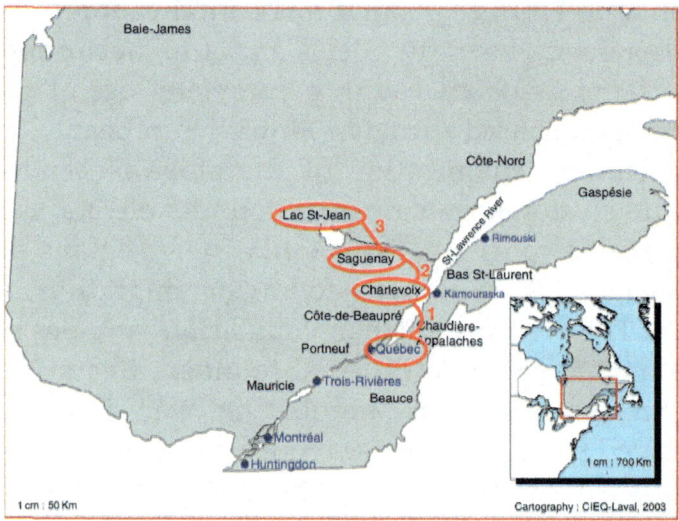

Figure 5-1. Serial founder events in the French-Canadian population along the St. Lawrence River from 1608 to 1852. Modified/reproduced with permission from Laberge A-M. et al. 2005. Population history and its impact on medical genetics in Quebec. Clinical Genetics 268: 287–301. Clinical Genetics.

This pattern of an initial founder effect (e.g., due to immigration from France) followed by serial colonizations provided conditions that allowed for random genetic drift to predominate over natural selection, especially along the expanding geographic wavefront. As a result, present-day French Canadians have been found to possess significant genetic differences compared to their source French populations. Genetic comparisons have revealed that French-Canadians along the expanding front have fewer variants than the Quebec populations from which they were derived. The Quebec populations, in turn, are genetically closer to the source French populations than those along the wavefront. Overall, it is estimated that the founder effect associated with the original population of 8500 settlers accounts for 80% of the genetic divergence of current French-Canadians from European-French populations, while the additional 20% has been generated by the subsequent serial founder effects. Consequently, there is reduced overall genetic diversity, a greater percentage of uncommon variants, higher levels of homozygosity, and only about 37% of genetic variants in common with European-French source populations.

Genetic consequences like these have significant implications for disease. Despite the reduced genetic diversity of French Canadians, the variants they harbor tend to be more damaging on average than their French source populations. This is particularly true regarding the persistence of harmful, less common variants from wavefront populations, which would tend to be eliminated more effectively by negative selection in larger core populations. The reduced number of genetic variants overall will increase the likelihood that mating pairs will each carry the same genetic disease allele at a given locus and pass them along to their offspring, making them homozygous for that allele. One study has found that wavefront individuals carry an average of 8.96 known pathogenic variants compared with 7.82 in their source populations, and the probability of homozygosity is 11.8% higher. This bears on the observation that French Canadians suffer from 22 or more Mendelian (single gene) diseases, some occurring at exceptionally high frequencies in more recently settled regions of Quebec. Some diseases with the greatest incidence include the absence of a brain structure called the corpus callosum, peripheral neuropathy, a condition affecting muscle movement (autosomal recessive spastic ataxia of Charlevoix-Saguenay), a congenital form of lactic acidosis (Leigh syndrome, French-Canadian type) and hereditary multiple intestinal atresias (abnormal obstructions).

Population Growth

Mutation, natural selection, and random genetic drift can be balanced when population size is stable. But the worldwide human "population" has been far from stable recently; rather, we have undergone exponential growth (Fig. 5-2). The global human population began to increase by approximately 0.5% per generation in the range of 10-25kya. In Europe, growth over the past 3,000-4,000 years has increased the effective population size 100-fold or more. Within the past two centuries, we have undergone an epidemiological transition with continued growth as high as 9% per generation, derived from massively reduced premature death rates. Even though natural selection is more effective in large populations, this growth has been possible mainly due to improvements in public health and medical care following the establishment of the germ theory of infectious disease and the discovery of antibiotics. These advances significantly diminished the impact of natural selection.

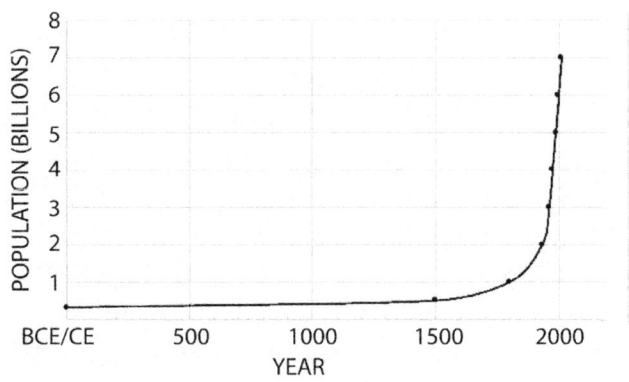

Figure 5-2. Worldwide human population size over the past 2000 years. Adapted from Roberts L. 2011. 9 billion? Science 333: 540-3.

Population growth has significant effects on our genome. Genetic simulations found that there is increased input of harmful mutations into the population, particularly rare mutations or ones occurring at low frequencies, resulting in an overall increase in genetic variation. A number of studies have confirmed these expectations. For example, a study out of the University of Washington examined over a million single nucleotide variants[16] from 6515 contemporary Americans[17]. They found that the number of variants was increased three-to five-fold compared with what would be expected for populations without growth, and there was an overabundance of rare variants. They estimated that 73% of all protein-coding single nucleotide variants arose in the last 5,000 to 10,000 years. Although this sounds like a long time ago, it is recent in evolutionary time and coincides with the population growth that began with the transition from hunter-gatherer to agricultural lifestyles. Among variants that were predicted to be harmful, about 86% arose during the same period. The rarest variants, comprising 55% of the total number, arose in the range of 1,000-2,000 years ago, very recently! Especially concerning is the finding that genes known to be associated with disease contained a higher proportion of recent harmful variants than other genes.

In summary, the dispersal from Africa and subsequent colonization were accompanied by widespread migration, bottlenecks, serial founder effects, and explosive population growth. The first three phenomena have acted to reduce local genetic diversity, such that the chromosomes of any contemporary individual from outside of Africa can be expected to carry much less genetic variation than individuals recently derived from Africa. At the same time, random effects and population growth have introduced many new mutations that increase the number of genetic variants. The combined effects of all the demographic processes are complex. However, the most consistent result seems to be an increased proportion of recently derived, rare, and often harmful mutations.

Sources:

Brace, Selina et al. 2022. Genomes from a medieval mass burial show Ashkenazi-associated hereditary diseases pre-date the 12th century. Current Biology 32(20): 4350-4359.

Carmi, Shai et. al. 2014. Sequencing an Ashkenazi reference panel supports population-targeted personal genomics and illuminates Jewish and European origins. Nature Communications 5:4835. https://doi.org/10.1038/ncomms5835.

Casals, F. et al. 2013. Whole-exome sequencing reveals a rapid change in the frequency of rare functional variants in a founding population of humans. PLoS Genetics 9(9): e1003815. https://doi.org/10.1371/journal.pgen.1003815.

Laberge A-M. et al. 2005. Population history and its impact on medical genetics in Quebec. Clinical Genetics 268: 287–301.

Skoglund, Pontus and Iain Mathieson 2018. Ancient genomics of modern humans: The first decade. Annual Review of Genomics and Human Genetics 19: 381-404.

SECTION TWO: CURRENT STATUS OF THE HUMAN GENOME

CHAPTER 5: Is Our Genome Perfect?

> *"None of us can cast stones, for we are all fellow mutants together."*[18]

> *"....as long as there is phenotypic variation, disease is inevitable...."*[19]

It should be clear from the preceding discussion that the answer to the question posed by this chapter is; no, our genome is not perfect, far from it! Many potentially harmful genetic variants can be found within the human genome. Some of these variants, particularly those that occur at low frequencies in the population, can have strong effects on both ecological fitness and the likelihood of disease. Even loss-of-function variants, which alter the structure of proteins so they can no longer perform their normal function, have been found at low frequencies in some populations. It seems reasonable to consider how such harmful mutations accumulate and persist in the human population when they are expected to be under strong purifying (i.e., negative) selection. To answer this question, it will be helpful to understand more about our genome.

The Human Genome

Our genetic heritage is what identifies each of us as belonging to the species *Homo sapiens*. At the same time, it also determines to a large extent, our unique individual characteristics as members of this species. We are not identical. For example, we all carry genes that determine our blood type; however, different base sequences within the DNA in those genes result in blood types A, B, or O. Differences in our genetic makeup can, to varying extents, determine our height, weight, eye color, hair color, skin color, personality traits, fertility, intelligence, vulnerability to disease and so much more! If our genes were identical from one person to another, natural selection could not lead to adaptation and evolution. This implies that genetic variations are associated with variable degrees of fitness, with some being clearly harmful. There are several reasons why potentially harmful ones persist, and each one of us carries hundreds of them. The presence of this kind of genetic variation can have significant consequences, especially under the shifting parameters of natural selection resulting from human culture, technology, and health care.

The human genome contains over three billion nucleotides and 63,494 genes, including nearly 20,000 protein-coding genes that compose only about 1% of the entire genome. Until fairly recently, we did not understand much about our genetic variability. The Human Genome Project was a collaborative international effort started in 1990 with the goal of determining all the base pairs that make up human DNA and identifying and mapping all the genes of the human genome. This massive effort was largely accomplished by 2003; however, a complete listing of every one of the 3,117,275,501 bases composing human DNA was not published until 2022[20].

But understanding the structure of "the human genome" falls far short of understanding the number and frequency of individual variations that combine to create the unique versions carried by each individual. To accomplish this, the specific genomes of very large numbers of individuals must be detailed and analyzed. Although this work has made significant progress over the past two decades and continues still, an enormous amount of effort is required to collect the data and perform the analyses. It would not be possible without significant technological advances that have allowed us to determine the sequence of DNA bases in large numbers of individuals. The original Sanger method (named after Frederick Sanger, its inventor) involves the creation of numerous fragments of different lengths from a source DNA. These fragments are then separated by size, and the bases are read by lasers in a device called a sequencer. Ultimately, reading increasingly larger fragments leads to the determination of the entire base sequence. Fortunately, since the completion of the human genome project, a variety of dramatically faster and less expensive techniques have been developed, collectively known as next-generation sequencing. Whereas the Sanger method provides for the sequencing of a single fragment of DNA at a time, next-generation sequencing can accommodate millions of fragments simultaneously. This advance, along with improvements in information processing, has enabled us to sequence and analyze more than 120,000 human genomes thus far, shedding light on the amounts and types of genetic variations in our populations.

When describing genetic variation, it is important

to understand the context. Some approaches reference global humanity, while others focus on specific populations of humans. Although there are millions of human single nucleotide variants, for example, many are rare and unique to specific populations, so they won't all be found within your own regional, ethnic, or cultural group. Even within a localized region, the genomes of individual persons will not necessarily reflect the variation present in the entire population. Large population studies typically find many rare variants, but any one person is unlikely to carry them; rather, most of an individual's variants are common, with only 1-4% having a frequency of less than 0.5%. Individuals can differ from each other by millions of genetic variants, although many of these are neutral and have no effect on function or health.

Much of our insight into human variability is derived from large studies such as the 1000 Genomes Project[21] and the whole-genome sequencing of 150,119 individuals from the UK Biobank[22], a dataset providing phenotypic information on hundreds of thousands of individuals across the UK. In the latter, more extensive study, 585,040,410 single nucleotide variants were found, along with 58,707,036 insertions/deletions and 895,055 structural variants[23]. The average individual carried nearly 8000 insertions/deletions. Humans possess a great deal of genetic variation! Overall, a typical individual may possess many times more variants in known regulatory regions of the genome than those that affect protein-coding. One of the most consistent findings among various studies is that human populations possess a greater number of rare or low-frequency variants than expected. In the UK study, 46%

of single nucleotide variants were found only in single individuals, and 96.6% have a frequency below 0.1%. To a large extent, this may result from the recent history of profound human population growth. These rare variants contain excess harmful mutations, which are often recessive. Among these, some had significant effects on phenotype, and new associations were found between rare variants and diseases.

The Ultimate Source Of Genetic Variation: Mutations

To better understand the origins of human genetic variation, we must take a closer look at the mutation process. Mutations reflect errors in copying DNA when cells reproduce. As we have seen, they often involve the replacement of one single nucleotide with another, leading to an alternate version of the DNA base sequence at the mutation site. Single nucleotide replacements do not always lead to a change in the genetic code; in those cases, the mutation has no effect and is neutral. As we have seen, many other mutations involve structural variants which can be inserted, deleted, or duplicated within the DNA strand. Even large portions of chromosomes or entire chromosomes can be gained or lost, such as trisomy 21 (three copies of chromosome 21) associated with Down Syndrome.

The most important mutations regarding heredity and evolution occur in the germline, where the egg and sperm cells are generated. The pairs of chromosomes in the germline cells and all the DNA they contain must first be duplicated, and a complete DNA copy must be forwarded into each of two new cells. Each of those cells then divides again without chromosome duplication, so only one chromosome of each pair ends up in each of the four new cells. These cells that contain half of the genetic content of the mature individual are called gametes (eggs or sperm). They also contain any mutations that may have been present in the germ cells or that occurred during the cell divisions. Mutations carried by the gametes are most important to heredity because they will be present in all the cells of the embryo following conception. That ultimately includes the germline cells of the next generation of adults, who may then pass them along to their progeny, and so on.

Individual mutations are rare events, although, in aggregate, they are surprisingly common. The question of how often mutations occur in the human germline has been the focus of much research and discussion since we obtained the ability to read DNA base sequences. There is no single mutation rate; rather, there is variability in the rate depending on the mutation type, the location in the genome, and many other factors. Two of the most significant factors are parental sex and age. Many studies show that most germline mutations are inherited from fathers, often attributed to the larger number of cell divisions needed to produce very large numbers of sperm over many years. It is estimated that 76% of all new mutations arise in the paternal germline. The rate of mutation also increases with paternal age and less so with maternal age. The number of new mutations transmitted paternally increases by about 1 to 2 per year, and offspring born to 40-year-old fathers harbor twice as many new mutations as the offspring of 20-year-olds. This has significant implications if reproduction is delayed, especially for men. Mutation rate can also be influenced by extrinsic factors such as exposure to radiation and chemical mutagens, both of which are likely to have increased in modern times.

New mutations are a significant ongoing source of genetic variation in human populations. It has been estimated that about 10 to the 11th power (the number 1 followed by 11 zeros - a very large number!) human germline mutations have occurred worldwide in the last generation and that a new mutation persists in a population for somewhere in the range of 25 to 167 generations. The average individual may be expected to carry about 50 to 100 new mutations at birth - each of which may be beneficial, harmful, or neutral with no effect.

About 1 to 10% of human mutations may affect fitness, and the average human carries well over a thousand harmful mutations with considerable variation even among healthy individuals. Most harmful mutations are likely to either have relatively subtle effects with only mild negative selection against them or to be recessive and only expressed in homozygotes so that they have less impact on fitness and can persist in populations. Mutations that have larger effects are more likely to be harmful with a negative impact on fitness.

DNA From Other Species

One surprising result from genetic studies over the past decade is the discovery that modern human genomes contain DNA sequences derived from related, archaic species: the Neanderthals and the Denisovans. These species colonized Eurasia at least 300,000 years before *Homo sapiens* arrived from Africa. Apparently, our ancestors interbred with these related species at least twice, raising the opportunity for archaic human DNA to mix with our own. Various studies have documented that around 2% of our present genome is derived from Neanderthal DNA, including genes that affect skin tone and hair color, sleeping patterns, mood, and smoking status. An important component of this DNA appears to influence the regulation of other genes. Both positive and negative selection have likely modified this genetic heritage, some of which has been linked to immunological, neuropsychiatric, and dermatologic disorders. There is also recent evidence that some Neanderthal genes affect the risk for COVID-19.

Natural Selection Does Not Result In Perfection

The ultimate effect of natural selection is to favor those genotypes that are best at perpetuating themselves, regardless of whether the favored genotype is, in any sense, perfect. Some genetic variants may have a harmful effect only in certain environments or when they co-occur with other variants, so they are not always subject to negative selection. An extreme example of this is sexual antagonism, where the same genes in different sexes may be subject to a different magnitude or even direction of selection[24]. In humans, there is evidence that mutations harmful to one sex may be beneficial to the other for traits such as height, weight, blood pressure, blood glucose, and total cholesterol. Recessive alleles may be sheltered from the impact of purifying selection when they co-occur with dominant alleles in heterozygotes. Genes that occur in proximity to one another on a chromosome are often said to be "linked" and, to some extent, inherited as a unit. As a result, some harmful mutations can become established by "hitchhiking" with beneficial ones to which they are genetically linked. Such harmful variants have been associated with autoimmune disorders, metabolic diseases, cancers, and mental disorders. An average individual is estimated to carry about 600 mildly deleterious mutations. Due to the weakness of purifying selection against such variants, some mutations can reach a high cumulative frequency in the human population through the process of random genetic drift.

Selection does not always act to reduce genetic diversity and, in some cases, may serve to promote it. As we have already seen regarding resistance to malaria, heterozygote advantage can serve to perpetuate alternate alleles in a population even when one of them is harmful in the homozygous condition. Diversity can also be maintained through frequency-dependent selection, where the advantage of a genotype decreases as it becomes more common.

Pleiotropy, when one gene or variant affects multiple traits, has been found to occur in about 17% of genes and about 5% of single nucleotide variants known to be associated with diseases. For example, a gene designated HLA-DRB1 is associated with both the production of immunoglobulin A (beneficial) and juvenile idiopathic arthritis (harmful). This type of situation, where a genetic variant produces two or more traits with opposing effects on fitness, is called *antagonistic pleiotropy*. Surprisingly, there are numerous examples where harmful genetic variants associated with human disease have been favored by natural selection due to their overall positive effects on fitness. We have seen this in sickle cell disease, where a hemoglobin variant impairs oxygen transport but protects against the malaria parasite. Another case involves a protein called huntingtin that is involved in several physiological processes and may play a role in long-term memory storage. However, one variant of the gene that codes for huntingtin produces an abnormal protein implicated in Huntington's Disease, which is characterized by progressive psychiatric disruption, cognitive deficits, and loss of motor coordination. Therefore, you would expect this variant to be eliminated over time by negative selection. However, the same variant has been found to have a significant selective advantage stemming from increased numbers of offspring in those who carry it. Although the underlying reason for this increased fertility is unknown, a Canadian study found that Huntington's patients had 39% more offspring than their unaffected siblings and 18% more than unrelated control patients. In addition, there is some evidence that Huntington's patients have a reduced cancer risk.

Other examples of antagonistic pleiotropy involve mutations in the APOE and BRCA1/2 genes. The APOE gene codes for apolipoprotein E, a protein involved in fat metabolism. A genetic variant of this gene is associated with improved performance in several different cognitive functions in the young but a greater risk of Alzheimer's or schizophrenia when older. Mutations in the BRCA1/2 genes in women cause an increased risk of ovarian and breast cancer. A study of women in Utah born before 1930 found that individuals with these mutations had more children, shorter interbirth intervals, and had their last child at a later age. At the same time, their post-reproductive mortality rate was increased.

There is evidence that genetic variants favored by natural selection for adaptation to cold environments can increase cancer risk later in life. In addition, geographic and environmental patterns of genetic risk for other diseases, such as biliary liver cirrhosis, alopecia areata, inflammatory bowel disease, systemic lupus erythematosus, and asthma, suggest links to environmental adaptation. However, the specific selective pressures are not known with certainty.

Some of these examples, where a mutation increases fitness earlier in life at the cost of increased disease or mortality later in life, also illustrate a phenomenon called mutation accumulation, a process underlying a popular theory of human aging. It postulates that genetically based disorders that occur later in life have less impact on reproduction and are, therefore, less subject to strong negative selection. As a consequence, harmful mutations that are expressed later in life will tend to accumulate. Moreover, the effect will be even more pronounced if those same mutations are associated with beneficial effects on fitness earlier in life.

Antagonistic pleiotropy also illustrates another important feature of evolution: tradeoffs. Not all traits can be simultaneously optimized, especially if they draw on common, limited energy reserves. Typical tradeoffs that occur in nature reflect the balance between reproduction and survival, offspring number and quality, or current versus future reproduction. As is true for the human genome, our bodies are not perfect or even optimally 'designed.'

Some of what we recognize as disease may represent complex trade-offs arising from conflicting adaptive pressures to increase fitness. The ability to walk on two legs rather than four provided better navigational potential for our hunter-gatherer ancestors. It freed up their hands for carrying items or using tools but also predisposed humans to hemorrhoids and back pain. During pregnancy, cephalopelvic disproportion reflects the situation where the birth canal is too small relative to the size of the baby's head. It is believed to have resulted from selection for both larger brain size at birth and a narrow pelvis to facilitate bipedal locomotion, and it commonly leads to the need for a Caesarian section in order to facilitate a safe delivery. Pregnancy also requires modulation of the maternal immune system so that the fetal tissues are not subject to immune rejection; simultaneously, the altered immune activity renders mothers more prone to some infections. Thus, some degree of harmful genetic variation affecting human traits can be maintained if there is an overall fitness benefit to the organism.

Many other harmful genetic variants may occur due to a mismatch between the current environment and different environmental conditions that favored them in the past. For example, the "thrifty genotype" hypothesis suggests that historical adaptation to an unreliable food supply may have favored metabolic changes that now predispose to non-insulin-dependent diabetes and other common systemic diseases in modern populations where food supply is abundant, and food quality is vastly different. Mutations in genes currently associated with cancer, coronary artery disease, and Alzheimer's disease may have been associated with fertility benefits in the past. It is also possible that many disease variants in modern populations may have originated as adaptively neutral mutations, which could have drifted over time to relatively high frequencies, but which then became harmful when subject to changing environmental conditions.

From the above considerations, we can understand that the "human genome" is not an identical collection of genes shared by all. There are alternate versions, or variants, of genes, and we all carry our unique set of these variants. Our individual collections of genes significantly affect our phenotype, or essentially, who we are, although our individual and collective genomes are not perfect. They harbor many potentially harmful genetic variations because natural selection does not necessarily lead to perfection.

Along with the population effects discussed in the previous chapter, harmful genetic variants can be perpetuated or even promoted. In some cases, their presence can manifest as disease. The next chapter will explore the relationship between genes and disease in more detail.

Sources:

Benton, Mary Lauren et al. 2021. The influence of evolutionary history on human health and disease. Nature Review Genetics 22: 269–283.

Byars, Sean G. and Konstantinos Voskarides 2020. Antagonistic pleiotropy in human disease. Journal of Molecular Evolution 88: 12–25.

Corbett, Stephen et al. 2018. The transition to modernity and chronic disease: mismatch and natural selection. Nature Reviews Genetics 19: 419-430.

Corona, Erik et al. 2010. Extreme evolutionary disparities seen in positive selection across seven complex diseases. PLoS ONE 5(8): e12236. https://doi.org/10.1371/journal.pone.0012236.

Fay, Justin C. 2013. Disease consequences of human adaptation. Applied and Translational Genomics 2: 42-47.

Goldmann J. M. et al. 2016. Parent-of-origin-specific signatures of de novo mutations. Nature Genetics 48: 935–939.

International Human Genome Consortium 2001. Initial sequencing and analysis of the human genome. Nature 409: 860-921.

Jacqueline, Camille et al. 2017. Cancer: A disease at the crossroads of trade-offs. Evolutionary Applications 10: 215–225.

Nurk, Sergey et al. 2022. The complete sequence of a human genome. Science 376(6588): 44-53.

Rodríguez, Juan Antonio et al. 2014. Integrating genomics into evolutionary medicine. Current Opinion in Genetics & Development 29: 97–102.

Rodríguez, Juan Antonio et al. 2017. Antagonistic pleiotropy and mutation accumulation influence human senescence and disease. Nature Ecology and Evolution 1: article number 0055. https://doi.org/10.1038/s41559-016-0055.

Shendure, Jay and Joshua M. Akey 2015. The origins, determinants, and consequences of human mutations. Science 349(6255): 1478-1483.

Shendure, Jay et al. 2017. DNA sequencing at 40: past, present and future. Nature 550: 345–353.

Sudmant, Peter H. et al. 2015. An integrated map of structural variation in 2,504 human genomes. Nature 526: 75–81.

Venter, J.C. et al. 2001. The sequence of the human genome. Science 291: 1304-1351.

CHAPTER 6: Genes, Fitness, And Disease

"For almost all human diseases, individual susceptibility is, to some degree, influenced by genetic variation."[25]

The concept of disease can be surprisingly tricky to define because the perception of disease depends on the context. It may be influenced by a mixture of social and economic factors, including economic class, gender, or ethnic group, and it may change over time. For our purposes, human disease can be defined as: any condition that causes pain, dysfunction, distress, social problems, or death to the person affected or similar problems for those in contact with the person[26]. Most of us will inherently know whether or not we are experiencing a disease.

In an evolutionary sense, is disease related to fitness? So far, we have tended to treat them as equivalent. We have described some genetic variants as harmful without specifying whether the harm applies to fitness, health, or both. They are, however, not equivalent. An organism's fitness reflects how well-adapted it is to its environment. It is enhanced by the organism's ability to survive and reproduce and is ultimately a function of how many of its genes are passed to future generations. Being fit in an evolutionary sense does not always require the organism to be in perfect health, free from suffering, or even be happy! In the absence of any direct effect on reproductive success, natural selection is indifferent to human well-being and suffering. Even if genetic changes that reduce fitness are associated with disease, they are only loosely related to its clinical severity. Consider, for example, a hypothetical new genetic mutation that virtually assures a person will acquire an untreatable form of pancreatic cancer at 70 years old and suffer a painful death. From a medical perspective, this is a horrible disease. However, 70 years is well beyond a human's usual reproductive lifespan. Unless the 70-year-old contributes substantially to the survival and subsequent reproduction of their progeny, there will be little to no effect of this mutation on fitness.

For any individual, the risk of contracting most diseases has a multifactorial basis and is often influenced by a complex interplay of genetic and environmental factors. As of January 2023, over 350,000 gene mutations have been identified that underlie or are closely associated with inherited human disease[27]. Even simple mutations, such as single nucleotide substitutions, can significantly impact biological functions. A typical individual may carry thousands of variants that affect the genome's protein-coding or regulatory regions. The impact of these mutations is quite variable, so only a fraction of individuals carrying them manifest disease. For this reason, a priority in medical genetics is to identify their biological effects and better characterize the extent to which they influence the risk of disease. It will be helpful to look at what we currently know about genes and disease in the modern world.

Characteristics Of Disease Genetics

In addition to single nucleotide substitutions, structural variants involving segments of DNA compose a significant portion of human genetic variation. For example, one study of the protein-coding portion of the genome revealed that a typical individual possessed 0.81 deleted genes and 1.75 genes that were duplicated, with most individuals carrying at least one rare copy number variant (where the number of copies of a specific segment of DNA containing up to thousands of bases varies among different individuals' genomes)[28]. Such variants have been found to play a role in diseases such as autism, bipolar disorder, and schizophrenia, while more common copy number variants are associated with susceptibility to cancer, infection, and metabolic disorders. One example of a specific disease caused by structural variations is Smith–Magenis syndrome, a developmental disorder characterized by intellectual disability, distinctive facial features, behavioral problems, and sleep disturbance. Most cases involve deleting a portion of chromosome 17 containing 3.7 million base pairs. Included in this DNA segment is a gene whose absence is believed to be responsible for the manifestations of the syndrome. Interestingly, a duplication of this DNA segment also results in a similar, yet distinctive condition called Potocki-Lupski syndrome. Structural variations also contribute to more complex, polygenic disorders (discussed below), such as attention deficit hyperactivity disorder, Crohn's disease, rheumatoid arthritis, type 1 and type 2 diabetes, autism, and schizophrenia.

The degree of *gene expression*, the rate and amount of protein product that a given gene produces, can vary widely among individuals. The human genome contains at least one million DNA sequences that encode instructions for the regulation of the expression of protein-coding genes. Most genetic variants that have been found to affect human traits appear to do so by affecting these regulatory sequences, and most human genes are under their influence. Many of them have greater than two-fold effects on the amount of gene product. Variation in regulatory sequences has a significant role in human adaptation but is also believed to be pervasive in common human diseases. For example, a mutation in a regulatory sequence influencing the expression of a gene designated as TBX5 has been found to cause congenital heart disease. In more complex disorders such as type 2 diabetes or cardiovascular disease, the exact mechanism by which regulatory variants affect the risk of disease can be quite complicated. In one case, a common single nucleotide variant is associated with the risk of myocardial infarction (heart attack). The less common allele at this site reduces the risk by upregulating a gene in the liver that produces a protein called sortilin. Increased production of sortilin, in turn, results in lower levels of low-density lipoprotein (LDL) cholesterol, which is a well-known risk factor for myocardial infarction.

A number of clinical syndromes result from mutations in mitochondrial DNA rather than nuclear DNA. Mitochondria are organized structures found in large numbers within most cells. They take in nutrients from the cell and break them down, producing energy that may then be used by the cell for a variety of functions. Mitochondrial diseases are not uncommon but are characterized by marked variations in their clinical features among patients. One example is blindness associated with a condition called Leber hereditary optic neuropathy, caused by mutations in any of four mitochondrial genes.

Simple Genetic Disease

Many diseases are inherited in a simple pattern consistent with a mutation in a single gene. This pattern of inheritance was first described in 1865 by a monk named Gregor Mendel, who performed experiments on inheritance in pea plants. Therefore, traits that are inherited in this fashion are often referred to as Mendelian traits. There are estimated to be 7,000 to 10,000 human diseases inherited in this manner. However, they are relatively rare, with recognized Mendelian disorders occurring in about 0.4% of live births. In addition, a substantial proportion of congenital disabilities likely have a Mendelian basis and are responsible for the deaths of 4,000 children annually in the US. Overall, over 25 million people in the United States are affected by simple genetic disorders. In addition to significant disability, suffering, and mortality, the economic burden for each child with a genetic disorder has been estimated to be $5,000,000 during their lifetime.

Cystic fibrosis is an example of a classic Mendelian disorder that is relatively well understood. It is a recessive disorder, so a person must inherit two copies of an abnormal gene variant to manifest the disease. It occurs in roughly one in every 3,000 births. People with cystic fibrosis suffer from wheezing or shortness of breath, persistent coughing, frequent lung infections, impaired growth, and abnormal bowel movements. The cause of this disorder can be one of several possible mutations in a gene coding for a protein called the cystic fibrosis transmembrane conductance regulator (CFTR). This protein normally regulates the proper flow of water and chloride in and out of cells lining the lungs and other organs, and CFTR derived from a gene with one of the cystic fibrosis mutations does not function properly. As a result, the secretions of cells that produce mucus, sweat, and digestive juices become sticky and thick. In the lung, mucus can clog the airways, trap germs, and increase the risk of infections. In the pancreas, mucus can interfere with the release of digestive enzymes, with adverse effects on the absorption of nutrients. CFTR modulators are one of the newest treatments for this disorder. These small molecules can partially restore the function of CFTR produced by some of the cystic fibrosis-associated mutations.

There are two ways in which individuals come to possess gene variants that cause Mendelian disease. The first way is to inherit variants already carried by one's parents. If the variant is dominant, as is the case for amyotrophic lateral sclerosis (ALS), the disease will be reliably inherited when that allele is passed to the offspring. More commonly, the parents may both be carriers of a recessive disease variant, as is the case for cystic fibrosis. This means that they each carry one copy of the disease variant along with one normal allele, so they experience the disease to a much milder degree, if at all. If the disease variant affects the production of an essential enzyme, for example, the normal copy of the gene may provide for enough enzyme to avoid any problems. The average healthy person carries single copies of about three recessive variants associated with severe pediatric disease, each accompanied by a normal copy of the affected gene (i.e., they are heterozygous). In the unfortunate circumstance where the offspring receive the same recessive variant from each parent, they will be homozygous at that locus and will suffer from the associated disease.

The other way in which individuals come to possess variants associated with disease is when the parents' genes are normal, but new mutations occur in the germline during the production of the egg or sperm that combine to form the new individual. New mutations that significantly disrupt gene function are believed to be the primary cause of rare sporadic malformations, such as Schinzel–Giedion syndrome, which is characterized by distinctive facial features, neurological problems, and organ and bone abnormalities. Most affected individuals do not survive past childhood.

Most Genetic Disease Is Not Simple

For any individual, the risk of contracting most diseases has a more complex basis and is more likely to be influenced by the interplay of genetic and environmental factors. In a complex disease, the presence of a particular gene variant in an individual contributes to the risk of disease but it is not enough to cause the disease. Common complex diseases include obesity, cardiovascular disorders, chronic obstructive pulmonary disease, type 2 diabetes, renal failure, and schizophrenia. In contrast to the progress made in identifying the specific genetic abnormalities underlying Mendelian disorders, it has proved to be more difficult to identify the specific genetic factors underlying the risk of complex diseases.

The most common approach for identifying genes involved in complex diseases has been the genome-wide association study (GWAS). These studies require massive amounts of data on large numbers of individuals. Usually, a specific human trait, often a disease, is the focus of any one study, and some determination regarding its presence or absence is noted for each person. At the same time, each person's DNA is subject to extensive analysis. A typical study examines one million or more single nucleotide variants. Statistical tests determine which ones occur more frequently in persons with the disease of interest. In one case, early GWAS's identified an association between a gene coding for a protein called Complement Factor H and age-related macular degeneration, a significant cause of blindness in the elderly. One variant of this gene increases the risk of macular degeneration by a factor of about 2 to 5 times and explains about 43% of cases in older adults[29]. As of 2021, more than 5,700 GWAS's had been performed for more than 3,300 traits. These studies identified over 50,000 positive associations between specific traits and genetic variants. Some diseases for which genetic risks have been identified include anorexia nervosa, major depressive disorder, cancers and subtypes of cancers, type 2 diabetes mellitus, coronary artery disease, schizophrenia, and inflammatory bowel disease.

Even early GWAS's clearly showed that large numbers of common genetic variants may be associated with common diseases, although the effect of each individual variant on risk tends to be small. An individual's genetic risk for a complex disease, therefore, depends in part on the cumulative effect of the particular combination of variants they carry. Such patterns of inheritance involving numerous genes are termed polygenic[30].

The contribution of rare variants (often defined by occurrence at less than 0.5% in the population) to human disease is more difficult to determine since they are not often identified in GWAS's. Why? Because they are rare! They may be confined to a single population, family, or even a single person. To find them, studies would require the inclusion of many more individuals than is usually the case. In an extensive, multi-institutional survey of protein-coding genes in over 15,000 individuals of European or African ancestry, over 500,000 single nucleotide variants were identified. Of these, 86% were rare, and 82% were population specific. Nearly 96% of the variants predicted to be functionally important were rare. It is estimated that in any one individual, 17% of variants are rare, while 65% are common, occurring at >5% on average.

Random genetic mutations with greater effects on the phenotype are often harmful, with negative effects on fitness. Therefore, natural selection should act to limit their frequency in populations to very low levels commensurate with their degree of harmfulness. Natural selection takes time, however, so there is a tendency for more recent and rare mutations to be more harmful. In GWAS's of disease variants, the youngest 20% account for almost four times the inherited risk than the oldest 20%. The overall contribution of rare variants to human disease continues to be explored. For example, one study of LDL cholesterol levels found an association with 157 mostly common variants, but a small number of additional rare or low-frequency variants affecting four genes accounted for as much as half of the heritability as all the common ones. There is also at least preliminary evidence that rare variants may significantly influence the risks of prostate cancer, type 1 diabetes, inflammatory bowel disease, autism spectrum disorder, and schizophrenia.

Gene variants are not the only factor contributing to the complexity of many diseases. The average heritability of human traits tends to be around 50%. This implies that genetics can account, on average, for about half the variation you see around you in complex traits such as height, weight, physique, or intelligence. There are clearly other influences that affect our characteristics. Prominent among these is the quality of our environment, including, for example, the availability of adequate nutrition or cognitive stimulation. The activity of DNA is sometimes modified by the addition of methyl groups or the binding of proteins called histones, and these alterations can have environmental triggers. They can be handed down over generations and sometimes play a role in human diseases. Gene-environment interactions occur when the effect of a gene is influenced by the organism's behavior or environment. For example, the impact of one allele on the risk of breast cancer depends upon how many children a woman has had, and another depends upon the woman's alcohol consumption. The effect of genes that render an individual more susceptible to an infectious disease depends upon exposure to that disease. Gene-gene interactions occur when the effects of one gene variant depend upon which variants of other genes are present. For example, suppose some gene functions in the human genome are duplicated. In that case, the effect of one gene impaired by a mutation may not be apparent since there is another gene that can compensate or lessen the impact. The inheritance of complex disease is complex!

All of the above considerations constitute the genetic architecture of a complex disease; that is, the total of all genetic influences contributing to that disease's risk. Specifically, it includes the number and frequency of relevant genetic variants in the population, the magnitude of their effects on disease risk, their relationship with alternative alleles at the same locus, how their function is modified by variants at regulatory sites, patterns of gene–gene and gene–environment effects, and pleiotropy. In addition, our increased insight into the genetic basis of human disease has led to a more "gene-centric" perspective of disease itself. Because of the genetic complexity underlying complex diseases, and the likelihood that various gene plus environment combinations may lead to the same manifestation of disease, some authors have argued that identifying common diseases categorized by symptoms is flawed. They suggest that such disorders simply represent the extremes of a continuum of traits and may have variable underlying genetic causes.

Genes And Disease: Diabetes As An Example

Diabetes has been one of the most extensively studied human diseases, and an overview of the underlying genetics illustrates several important features of the role of genes in disease risk. Insulin is a hormone produced by β cells in the pancreas that helps regulate the amount of glucose in the blood. Diabetes occurs when the body's insulin-mediated response to elevated blood sugar levels is disrupted. We will look at four different clinical categories of this disease: neonatal, maturity-onset, type

1, and type 2 diabetes.

The first two types of diabetes are examples of simple genetic disorders. Neonatal diabetes is a rare, severe disease presenting during the first six months of life. It results from a mutation in one of two genes involved in the synthesis and secretion of insulin. As a result of this mutation insulin production will be inadequate, and blood sugar levels will be abnormally high, resulting in immediate and long-term threats to health. Maturity-onset diabetes of the young is another form of the disease, accounting for 1-2% of all diabetes cases. It may result from a mutation in any one of a dozen or more genes.

Type 1 diabetes is an autoimmune disorder that characteristically occurs early in life. Autoimmune disorders result when the body's immune system becomes abnormally activated to attack normal body tissues as if they were invading organisms. In the case of type 1 diabetes, the target is the insulin-producing cells of the pancreas. The result is severe insulin deficiency and elevated blood glucose, which may lead to diabetic coma and long-term consequences such as increased susceptibility to infection, cardiovascular disease, and damage to the nervous system.

The risk of type 1 diabetes is affected by both genetic factors and the presence of environmental triggers, such as pathogens or toxins that activate the immune system. The genetic component accounts for about half of the disease risk, and most of that risk can be accounted for by over 60 known gene variants identified in genome-wide association studies. The contribution to the overall risk for diabetes from these individual variants is quite variable but is generally modest. The most significant variants, accounting for up to 30-50% of genetic risk, are alleles related to human leukocyte antigen (HLA) class II genes. These genes are involved in the processes by which the immune system distinguishes the body's own proteins from proteins made by foreign invaders such as viruses and bacteria. One specific combination of two of these risk variants can increase the risk of diabetes 16 to 63 fold! Almost half of the children with this genetic combination will develop antibodies to pancreatic β cells before the age of 5 years. Other variants outside of the HLA complex, such as those affecting the insulin gene itself, can also increase risk. Despite the importance of genetics in determining the risk of diabetes, the likelihood of identical twins both developing type 1 diabetes is only in the range of 40 to 60%, even though they share the same genetic variants. Factors such as environmental triggers and lifestyle also play a significant role in this type of diabetes.

By far, the most common form of diabetes is type 2, accounting for up to 95% of people with diabetes. Unlike other forms, the onset is usually in mid-life or later. Patients with type 2 diabetes have a resistance to the effects of insulin rather than an insulin deficiency. As a result, various body tissues fail to respond appropriately to normal insulin levels, and the pancreas cannot produce enough insulin to compensate and prevent the elevation of glucose levels.

There is a genetic component to type 2 diabetes, but it is complex and challenging to quantify. Over 400 genetic variants affecting the risk have been identified. They are common and involved in regulating insulin secretion or sensitivity, with low to moderate effects. Collectively they account for only about one-fifth of the genetic component of risk, although individuals who carry the greatest number of risk alleles have a nine-fold increased prevalence of the disease compared with groups that carry the least number of risk alleles. There are also 24 known rare variants and 56 low-frequency variants. Although some of them have significant effects, they do not contribute much to overall risk compared to the numerous, more common variants of lesser effect. Changes in the regulation of gene activity due to environmental factors are also likely to influence type 2 diabetes risk.

The different forms of diabetes demonstrate various aspects of the genetic basis of disease. Some forms can be attributed to variations in a single gene (neonatal diabetes and maturity-onset diabetes of the young), another is polygenic with a significant role for rare variants with large effect (type 1), and one is highly polygenic with most risk derived from common variants of small to modest effect (type 2). The example of diabetes also underscores the importance of the environment and gene-gene interactions in determining the overall disease risk.

Sources:

Albert, Frank W. and Leonid Kruglyak 2015. The role of regulatory variation in complex traits and disease. Nature Reviews Genetics 16: 197-212.

Bomba, Lorenzo, Klaudia Walter and Nicole Soranzo 2017. The impact of rare and low-frequency genetic variants in common disease. Genome Biology 18: 77. https://doi.org/10.1186/s13059-017-1212-4.

Chong, Jessica X. et al. 2015. The genetic basis of mendelian phenotypes: Discoveries, challenges, and opportunities. The American Journal of Human Genetics 97: 199–215.

Claussnitzer, Melina et al. 2020. A brief history of human disease genetics. Nature 577: 179-189.

Edwards, Albert O. et al. 2005. Complement Factor H polymorphism and age-related macular degeneration. Science 308: 421-424.

Fu, Wenqing, Timothy D O'Connor and Joshua M Akey 2013. Genetic architecture of quantitative traits and complex diseases. Current Opinion in Genetics & Development 23: 678–683.

Fuchsberger, C. et al. 2016. The genetic architecture of type 2 diabetes. Nature 536: 41–47.

Giwa, A. M. et al. 2020. Current understandings of the pathogenesis of type 1 diabetes: Genetics to environment. World Journal of Diabetes 11(1): 1-25.

Ingelsson, Erik and Mark I. McCarthy 2018. Human genetics of obesity and type 2 diabetes mellitus. Past, present, and future. Circulation: Genomic and Precision Medicine 11: e002090. https://doi.org/10.1161/CIRCGEN.118.002090.

Nickels, Stefan et al. 2013. Evidence of gene–environment Interactions between common breast cancer susceptibility loci and established environmental risk factors. PLOS Genetics 9(3): e1003284. https://doi.org/10.1371/journal.pgen.1003284.

Tam, Vivian et al. 2019. Benefits and limitations of genome- wide association studies. Nature Reviews Genetics 20: 467–484.

Tennessen, Jacob A. et al. 2012. Evolution and functional impact of rare coding variation from deep sequencing of human exomes. Science 337: 64-69.

Uffelmann, Emil et al. 2021. Genome-wide association studies. Nature Reviews Methods Primers 1: 59. https://doi.org/10.1038/s43586-021-00056-9

Weischenfeldt, Joachim et al. 2013. Phenotypic impact of genomic structural variation: insights from and for human disease. Nature Reviews Genetics 14: 125-138.

SECTION THREE: RELAXATION OF NATURAL SELECTION

CHAPTER 7: Human Culture And Natural Selection

"The evolutionary uniqueness of man lies in that in mankind the biological evolution has transcended itself"[31]

Many potentially harmful genetic variants can be found within the human genome. Some of these variants, particularly those that occur at lower frequencies in the population, can have fairly strong effects on both ecological fitness and the likelihood of disease. However, natural selection plays a significant role in limiting the transmission of these variants across generations. In this chapter, we are going to look at recent cultural advances in our species and their effects on natural selection.

The Role Of Culture In Human Evolution

A unique feature of evolution in humans compared to other organisms is the profound impact of human culture. What is culture, exactly? In general terms, it can be defined as information that is socially transmitted. The information can be wide-ranging: we learn and share our religious beliefs, national traditions such as the observance of Thanksgiving in the US, medical therapies, and information about how to construct nuclear power plants. The degree to which culture and cooperation have developed in humans has allowed us to create technology, social institutions, knowledge systems, and behavioral practices to a degree far beyond the capacity and complexity of any other species' socially learned behaviors.

Although advanced intelligence and the capacity for development of a complex culture is a biological characteristic of the human species, the specific content of human culture is not innate or coded in our genes. Nevertheless, the fact that there is variation in the content of culture over space and time, combined with the ability to pass it along from one generation to the next, provides the potential for cultural adaptation and a type of evolution analogous to genetically based Darwinian evolution. But there are also profound differences. Genetic changes are largely random, often not beneficial from an adaptive standpoint, and certainly not goal oriented. In contrast, changes in culture are often intentional and goal-oriented.

One of the most significant differences between genetic and cultural evolution lies in the far greater rate and flexibility of transmission of the latter. Technological advances have provided almost unlimited capacity for storing cultural content outside of the human brain and body, with increasingly efficient mechanisms for storage and retrieval. There is no genetic evolutionary equivalent to Google! Cultural content can also be replicated on massive scales and disseminated rapidly, in theory, to the entire human population. Consequently, transmission occurs not just vertically (between generations) but also horizontally (over space within time frames much shorter than a generation). This provides a remarkable potential for ultra-fast adaptation, even within individual lifetimes: many of us who grew up without computers now use them routinely in nearly all aspects of our lives. The combination of human intelligence and cultural adaptation/evolution has been a major factor facilitating the rapid expansion and occupation of widely divergent environments throughout the globe, far more quickly than would have been possible based on genetic inheritance and Darwinian evolution alone.

Whereas other organisms primarily evolve genetically in order to adapt to their environments, we have evolved culturally to adapt our environments to suit our needs. In ecological terms, humans have become masters of "niche construction." An ecological niche is the unique role and position a species has in its environment; how it meets its needs for food and shelter, survives, reproduces, and avoids competition with other species. Most species succeed ecologically by adapting to their environments in a specialized way, mainly through genetic evolution. Some birds, for example, have evolved bills that are structurally adapted to eat specific kinds of fruit or nuts. But humans have succeeded beyond measure primarily by modifying our environment to accommodate our existing genetic constitution. As a result, we have actively constructed niches that did not previously exist. Some of the most significant examples of our cultural evolution are cited in the following table.

Age	Cultural Features
STONE AGE	
Paleolithic Ending 10,000 B.C.	• Stone/bone tools • Hunter/gatherers • Cooked prey using fire • Caves or simple huts
Mesolithic 10,000-8,000 B.C.	• Small stone tools • Spears and arrows
Neolithic 8,000-3,000 B.C.	• Pottery • Agriculture • Sewing/weaving • Home construction • Art
BRONZE AGE 3,000-1,300 B.C.	• Metalwork replaces stone weapons, tools • Metal plows/wheels • Textiles • Warfare • Religion • Earliest writing • Government/law • Roundhouses • Villages/cities • Advances in architecture/art
IRON AGE 1,300-900 B.C.	• Iron technology • Steel tools/weapons • Advances in religion, writing systems, documentation • Larger homes, stables, forts, palaces, temples, religious structures • City planning • Advances in agriculture/art
SOME RECENT ADVANCES	• Secure dwellings with climate control • Agricultural technology • Sanitation practices • Improved nutrition • Water purification • Pharmaceuticals • Pasteurization • Refrigeration • Labor-saving technology • Surgical and medical therapies • Sterile practice in medicine • Vaccinations • Sensory restoration, e.g., eyeglasses, hearing aids • Neuropsychiatric care • Physical rehabilitation • Supportive services, e.g., non-profits, Medicare, Medicaid

Table 8-1. Major advances in human culture. Partially adapted from Kennedy, L. 2019. The Prehistoric Ages: How Humans Lived Before

Written Records. https://www.history.com/news/prehistoric-ages-timeline. Accessed January 19, 2023.

    Although some experts have gone so far as to suggest that human genetic evolution has ceased because of our profound ability for cultural adaptation, this is not the case. Our capacity to modify our environments can create novel selective pressures that may, in some instances, serve to counteract selection that would occur in its absence. Our constructed niche has the potential to favor traits whose contribution to fitness depends upon a culturally altered environment. A well-documented example is the genetically based prolongation of functional lactase activity extending into adulthood following the cultural development of pastoralism, as we discussed previously in Chapter 3. Our ability to modify our environment appears to have resulted in especially strong selective pressures, having major effects on recent human genetic evolution and potentially leading to the perpetuation of alleles that would otherwise be harmful[32]. There may be a coevolutionary relationship between genes and culture, each affecting the other. It is possible that the uniqueness of *Homo sapiens* arose because of increasing dependence on our (genetically based) cognitive abilities, which, in turn, improved our ability to modify our environments, leading to further cognitive development and so on. Human evolution continues this way but in response to a significantly altered set of selection pressures.

The Relaxation Of Natural Selection

Relaxed selection refers to the reduction or elimination of a source of selection that was formerly important during the evolution of a species. As environments change, genetically based traits that were previously adaptive and subject to positive or purifying selection may lose their adaptive advantage. Selection favoring these traits will then be said to be "relaxed." In the absence of purifying selection to maintain a trait, the genetic basis of that trait may degenerate over time due to random changes in the frequency of alternative genotypes. If there are metabolic or adaptive costs in maintaining that trait, it may even become subject to negative selection and ultimately be eliminated from the population.

This concept can be illustrated in a well-documented example from nature involving the Mexican cavefish *Astyanax mexicanus*. This species is believed to have diverged into different forms around 10,000 years ago. The original form lives on the water's surface and has eyes, while cave-dwelling forms have no eyes. There are two schools of thought on how the loss of eyes might have occurred. One idea is that natural selection for maintaining eyesight is lost in the dark environment of the caves, and over time random accumulation of mutations in the genes involved in eye development has compromised their function. The other school suggests adaptive advantages to eliminating the eyes, such as reallocating the resources devoted to eye development to other more adaptive functions. Following a detailed analysis of the situation, one biologist suggests the explanation involves pleiotropy[33]. It is possible that genetic variants underlying the loss of eyes during development have been favored by natural selection due to their positive effects on other traits that promote fitness in the cave environment. Whatever the cause, if cave-dwelling, eyeless fish were re-introduced to the surface aquatic environment, they would likely be at a severe adaptive disadvantage.

To a certain extent, the environments – physical, social, and medical – that humans have created for ourselves are analogous to the relatively protected environments within which captive non-human species are bred. This occurs, for example, when endangered species such as the California Condor are bred in captivity to supplement the few remaining individuals in their natural environment. Like humans, these animals may be protected from predation and physical threats, provided abundant food and opportunities for mating, and medical care to prevent or treat illness. Under these circumstances, natural selection is relaxed, altering the mutation-selection balance in favor of allowing harmful variants in the population to persist. There may even be inadvertent selection for individuals that thrive in captivity but could be disadvantaged in their natural, wild conditions. Simulations demonstrate that variation in typical traits of these species drastically increases within a few generations due to the absence of purifying selection[34]. High trait variance can help populations adapt to changing conditions, but it is not necessarily good for the individuals! Some of this variance is likely derived from alleles that would be harmful in the natural environment. When these animals are released back into the wild, natural selection is restored and will once again promote adaptation to the natural environment, but many individuals may die or fail to reproduce in the process. As a result, the variance acquired in captivity is rapidly lost, and the population size is reduced. To illustrate this point with a hypothetical example from nature, consider a small mammal on a remote island with no predators, although predators live on other surrounding islands. In the absence of selection pressure from predators, variability would be expected to increase in any genes that affect behaviors related to avoiding predators, rendering this population potentially more vulnerable. These populations would likely be decimated should their predators be introduced, producing strong selective pressure for predator avoidance. Those who do not succeed in this will not contribute to subsequent generations.

In many ways, human niche construction and cultural evolution have taken precedence over genetic evolution, leading to the relaxation of the selection pressures that historically molded our species. In the following chapters, we will explore the implications for the future well-being of both individuals and their societies.

Sources:

Dobzhansky, Theodosius and Gordon Allen 1956. Does natural selection continue to operate in modern mankind? American Anthropologist 58: 591-604.

Laland, Kevin N. et al. 2010. How culture shaped the human genome: bringing genetics and the human sciences together. Nature Reviews Genetics 11: 137-148.

Rinaldi, Andrea 2017. We're on a road to nowhere. Culture and adaptation to the environment are driving human evolution, but the destination of this journey is unpredictable. EMBO Reports 18: 2094-2100.

Ross, Cody T and Peter J Richerson 2014. New frontiers in the study of human cultural and genetic evolution. Current Opinion in Genetics & Development 29: 103–109.

Taylor, Timothy 2010. The Artificial Ape: How Technology Changed the Course of Human Evolution. St. Martin's Press: New York.

Zampieri, F. 2017. The Impact of Modern Medicine on Human Evolution. In: Tibayrenc, Michel and Francisco J. Ayala, eds. On Human Nature. Biology, Psychology, Ethics, Politics, and Religion. Elsevier Science: New York.

SECTION FOUR: IMPLICATIONS

CHAPTER 8: Consequences Of Relaxed Selection

Relaxed Selection In Humans

Historically, physical conflict, famine, and infectious disease were significant causes of death and agents of natural selection in humans. They served to shape, optimize, and maintain the human genome. Early death, maternal mortality, and failure to reproduce successfully were common until relatively recently. In stable populations, an average of only two of the five to eight children typically produced by an adult female survived to reproductive age. Infectious diseases such as pneumonia, gastroenteritis, malaria, and measles were major sources of mortality during childhood, killing half of all children by age 15 and contributing to an average life expectancy of around 20 years. The Black Death alone is thought to have killed about a third of the European population, and as recently as the end of the 19th century, only 35% of Europeans reached the age of 40. Genetically based variants that made us more susceptible to any of the varied causes of premature death or reproductive failure would tend to be eliminated from the population by the ongoing process of negative selection.

In recent times, all of that has changed. The practices of human niche construction - including modern medicine, improved nutrition, and public health practices - have profoundly affected survival and reproduction by modifying the most historically important agents of natural selection. In many developed countries, the opportunity for selection by differential mortality has dropped precipitously, e.g., by a factor of 10 in the United States over the past century. Fetal mortality has declined by approximately 99% in England since the 1500s, while maternal mortality has decreased to as low as one-fiftieth the rate that prevailed in the early 20th century. Over the past two centuries, human life expectancy has increased by about 2.5 years per decade. Reducing childhood poverty and disease in developing countries has been, and continues to be, a global health priority leading to a global decline in childhood mortality of more than 2% per year between 1990 and 2010. The death rate before age five was nearly cut in half over the past two decades ending in 2019. At the same time, new medical treatments have the potential to restore fertility to persons whose genetic makeup may have previously been associated with reproductive failure. For example, 98% of persons born in Australia now have a full opportunity to reproduce. A publication from the Institute of Evolutionary Medicine at the University of Zurich has concluded, "For the first time in the evolution of humanity, the majority of natural selection pressures were relaxed to the apparent benefit of all of us"[35].

No one can deny that these profound changes reflect equally profound improvements in the quality of our lives and longevity. The evolution of human intellect and cultural adaptation has provided countless opportunities to enhance our lives in ways that are not possible for other species. We cherish the gains we have made, and our religious and ethical beliefs promote the concept of helping to improve the lives of the less fortunate among us. All of this is a good thing for humanity. But there are significant differences between what is good for us as individuals versus what is best for our species in the long term. We need to look no further than the climate crisis for an example of how cultural adaptations, such as the internal combustion engine, can enhance our lives while threatening the well-being of our species (and all life on the planet) in the long term. It is ultimately a matter of perspective.

There is a dark side to our achievements as a species that is not so obvious but could ultimately threaten our quality of life, social structures, and long-term success. Historically, natural selection has provided a mechanism for eliminating harmful variants and maintaining optimal genotypes, in addition to the proliferation of new adaptations. However, the parameters of natural selection have been dramatically altered compared with the forces that initially shaped our species. The cost of cultural adaptation may well be a gradual degradation of the very genetic blueprint that makes us who we are.

Before discussing our future genetic challenges, let's review what we have learned about our current genetic lineage. There is ample evidence that the human genome has undergone extensive adaptive evolution, even in relatively recent times. Three other processes have also had a profound impact: a history of numerous bottlenecks and serial founder effects, continuing cultural evolution with the relaxation of historical selection pressures, and recent and ongoing profound population growth (itself a result of relaxed selection). The combined effect of these processes has led to a genome characterized by locally reduced genetic diversity (outside of Africa) and an overabundance of low-frequency or rare genetic variants, many of which are potentially harmful. Some harmful variants have always been maintained in our populations due to genetic phenomena such as heterozygote advantage, antagonistic pleiotropy, and other evolutionary tradeoffs. Previously, in more stable populations, the rate of new mutations was balanced by purifying selection that acted to remove harmful variants. But contemporary human environments provide an "evolutionary override" that buffers our genome from ancient selective pressures that served to limit the transmission of harmful mutations, and alters the processes that promote ongoing evolution.

Genetic Concerns

Given our society's focus on individual health, well-being, longevity, and reproductive success, the potential genetic cost of relaxed selection for the human species is considerable. The accumulation of random mutations over time may lead to increased genetic variability and the risk of anatomical or functional deterioration of formerly adaptive traits. Individuals with a broader range of beneficial and potentially harmful genetic variants now contribute to subsequent generations, thus perpetuating their genetic makeup. It has been noted that disorders such as visual impairment, color blindness, and many anatomical variations or congenital abnormalities occur less frequently in more primitive human populations than in advanced cultures, where their negative impact on fitness is largely ameliorated. Cultural adaptations that diminish the negative impact of genetic variants associated with such disorders raise the likelihood of perpetuating those variants, thereby increasing genetic risk. From an evolutionary standpoint, it is not necessarily bad for a species to acquire a certain amount of genetic variability. It is such variability that allows populations to adapt more quickly to changes in selection. As we have seen, though, the cost of adaptation at the population level is death and reproductive failure at the individual level.

Concerns about the consequences of relaxed selection and an increased load of mutations on our species began to be expressed in the late 19th century, long before our current understanding of human evolutionary genetics. They were explicitly recognized in the World Health Organization's report on the status of the prevention and treatment of genetic disorders in 1972:

> ...*empirical observation indicates...that the incidence of pyloric stenosis* [a condition that blocks food in the stomach from entering the small intestine of newborns] *in the offspring of treated patients* [i.e., patients who themselves had been surgically treated for pyloric stenosis] *is about 5%* [compared with 0.24% in Caucasians without such a history]...*The incidence of affected children among the offspring of the now numerous survivors of* [pediatric] *cardiac surgery is approximately 3% in comparison with an incidence at birth of about six per thousand* [0.6%] *in the general population. The incidence of affected offspring among surgically treated patients with myelomeningocele* [a neurological malformation]...*is also likely to*

be of the order of 3 to 4% [compared with 0.06% in persons without a personal history of myelomeningocele]. *Thus, it may be anticipated that over the next few generations, the incidence of these multifactorial traits will increase by 3 to 5% per generation, provided that there is no change in the environmental factors contributing to the disease[36].*

Childhood disorders that can be surgically corrected, such as pyloric stenosis, are prime candidates for increased genetic risk over generations due to the acquired ability to survive to reproductive age. More recently, relaxation of selection has been implicated in the increased incidence of deafness related to recessive mutations in a protein-coding gene and the increased prevalence of diabetes mellitus type 1. In general, however, the genetic consequences of relaxed selection have been overlooked as a contributing factor to the increasing burden of many diseases in contemporary society.

The impact of changes in simple genetic traits should be most apparent, but harmful genetic variants that may pose the most significant difficulty will likely involve complex polygenic traits that can contribute subtly to overall genetic deterioration. The effects may not be readily evident due to cultural compensation but could involve maladies that do not currently impact survival and reproduction or even qualify as "diseases." For example, consider individuals who have greater than average blood sugar levels but do not meet the criteria for diabetes. Others may have poor tolerance for sustained physical activity or fall in the lower ranges of intelligence. Some of us suffer from frequent colds or experience frequent headaches. Others may have anxiety, suffer sleeplessness, have reduced hand/eye coordination, or have a myriad of other such traits. Michael Lynch, a population geneticist at Arizona State University, has suggested that "obvious candidate characters for monitoring include problems with visual acuity, tooth alignment, child-birth difficulties, and infertility, all of which historically must've been subject to strong selection, but which are now greatly modified by medical intervention[37]."

Human disease burden is known to be increasing worldwide, particularly for chronic conditions which are complex and polygenic. Chronic diseases with significant impact include cancer, diabetes, hypertension, stroke, heart disease, respiratory diseases, arthritis, obesity, oral diseases, depression, and back/neck pain. These ailments can markedly reduce one's quality of life and/or lead to premature death. They impact our societies due to recurrent hospitalizations, the need for extensive medical care, and long-term disability. They are also among the most prevalent and costly health conditions in the United States. Nearly half of all Americans are estimated to suffer from at least one chronic disease, and one in four have two or more such conditions. In children, obesity, asthma, diabetes, and learning/behavior problems are increasing. Globally, rates of all metabolic diseases (hypertension, type 2 diabetes mellitus, hyperlipidemia, obesity, and non-alcoholic fatty liver disease) increased from 2000 to 2019, particularly in more socio-economically developed countries[38]. Rates of type 2 diabetes are skyrocketing. Over 100 million adults in the US have diabetes or pre-diabetes, and it is estimated that it will affect 700 million people globally by 2045. In addition, early-onset cancers in adults younger than 50 are increasing[39]. It is well recognized that many of these disorders are strongly influenced by behavior and lifestyle and, at the same time, are heritable, indicating that there are likely to be significant contributions from both genes and the environment. This means that the risk imposed by lifestyle will vary with the individual's genetic background. The causes of increasing morbidity in our society are multifactorial and complex. Still, the potential role of relaxed selection in accounting for an increased genetic burden of disease risk has largely gone unrecognized.

This may be starting to change. One study published in 2018 examined trends in obesity, a genetically complex trait associated with premature mortality and several common, chronic diseases[40]. It has tripled in prevalence over the past four decades and, as of 2016, affects 12% of the world's adult population, posing a major public health concern. Variants in over 1000 genes have been linked to obesity, with heritability in the 40 to 70% range. The 2018 study examined trends in the body mass of young Swiss males over a period of 140 years ending in 2012. The authors analyzed data on nearly 60,000 male conscripts (compulsory in Switzerland) aged 18-19 years and found that both weight and body mass index (BMI) of the new classes of conscripts increased over time[41]. More interesting is the finding that variability in weight and BMI also increased, with greater deviation from the average at both the upper and lower ends of the weight spectrum. This increased variability was accompanied by increases in blood markers of inflammation at both extremes. It is clear that obesity is affected by both environmental and genetic factors. The authors identify a potential evolutionary effect, with relaxed selection leading to increased genetic variability in body composition and metabolism, posing significant health risks. In the same year, a second study also reported a link between relaxed selection and obesity, with a stronger association in males than females[42].

The Biological State Index (BSI) is a statistical measure based on age-related mortality and reproductive potential that reflects the strength of natural selection acting upon a population. It can also be interpreted as the probability for an average person in a population to have an opportunity to pass their genes to the next generation. The value of the BSI ranges from zero; with no chance of the average individual contributing to the next generation, to one; where reproductive potential is achieved with minimal effect from mortality and, therefore, little opportunity for natural selection. Using this statistic applied to historical data from Australia, it was found that the opportunity for natural selection has decreased markedly over the past 120 years, with most individuals able to pass along their genes to subsequent generations (Figure 9-1). Globally, the opportunity for natural selection has decreased by about 2.5 times over the same period.

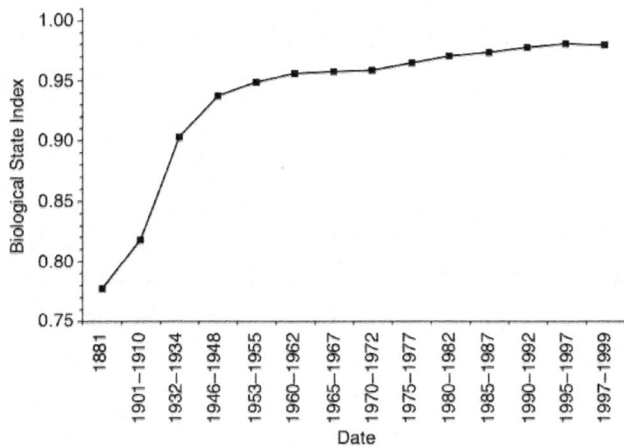

Figure 9-1. The Biological State Index over time in Australia. Reprinted from Medical Hypotheses 57(5): Stephan, C. N. and M. Henneberg. Medicine may be reducing the human capacity to survive, p. 634, copyright 2001, with permission from Elsevier.

Wen-Peng You from the University of Adelaide and Maciej Henneberg from the University of Zurich have published a series of reports employing the BSI to explore the possible role of relaxed natural selection in human disease[43]. Using vital statistics from 118 to 204 countries (depending upon the study), they found significant correlations between reduced natural selection and increased incidence of type 1 diabetes, cancer, and dementia. For cancers that are known to be strongly genetically based, the incidence rates in the ten countries with the least opportunity for selection by mortality are 5.7 times greater than the incidence rates for the ten countries with the greatest opportunity for selection. The relationship held for younger age groups and not just the elderly for cancer and dementia. There is no doubt that other environmental changes in more developed countries could contribute to these findings, but even when the authors accounted for factors such as the gross domestic product (GDP), life expectancy, sugar consumption, obesity, physical inactivity, smoking and urbanization, the correlations of disease with reduced selection remained.

Dr. Ivan Fuchs, an Israeli psychiatrist, published a book in 2019 that details the possible role of relaxed selection in increasing the incidence of human mental and behavioral disorders[44]. He argues that the relaxation of our past selection pressures affecting behavior leads to excessive genetic diversification, extends the extremes of normal human behavior, and increases the predisposition to mental disorders.

We are facing genetic challenges in the 21st century. Understanding the time frame within which we will likely see increased disease susceptibility due to the accumulation of harmful mutations is very helpful. The proportion of harmful single nucleotide variants has been shown to have increased in recent human history, accounting for up to 48% of all variants within a single genome. The rate at which new mutations enter the population can vary considerably based on several factors, including paternal age and exposure to environmental mutagens, both of which are likely increasing in more "developed" societies. It's even possible that the rate of new mutations could increase when selection is relaxed, as the mutation rate itself may be limited under the influence of natural selection. It is difficult to make any predictions with certainty; the evolving genetic architecture of human traits results from the complex interplay among genes and their expression, pleiotropy, the environment, gene-gene, and gene-environment interactions. Nevertheless, along with the 3-5% increase in harmful traits per generation cited by the WHO (above), additional estimates have been offered by other authors[45]:

- Loss of fitness of 1-5%/generation
- Decrease in overall viability around 2%/generation
- Mendelian recessive diseases increase 10%/100 years
- Decline in "genetic quality" of 7.7%/generation

It is clear that the adverse effects of relaxed natural selection can be seen within time frames as short as one human generation. It has also been predicted that the mean phenotypes of the residents of industrialized nations are likely to be somewhat different in just two or three centuries, with significant impairment in morphological, physiological, and neurobiological functions. Ultimately, the accumulation of harmful variants in advanced societies can have a long-term impact on the human gene pool. As expressed by the American Nobel prize-winning geneticist Hermann J. Muller in 1950, "The frequency of mutant genes must rise, with resultant degeneration of the biological organism[46]."

Implications For Society

It is apparent that what is good for a single individual's health may not necessarily be good for society, or humankind in the long term. In fact, the loss of optimal genetics in humans may have profound consequences for society. Increases in genetic disorders or diseases will likely amplify human suffering and disability. To maintain the quality of life and longevity to which we are accustomed, we will need to rely more and more heavily upon existing and evolving medical technologies and social support.

While new therapies and technologies offer the potential for improved management of medical disorders, the economic costs are often considerable, presenting an increased burden to societies struggling to provide affordable care to all their citizens. If you have been prescribed a "new and improved" medicine lately that is still under patent, you have already felt the impact of increasing drug costs. The pharmaceutical industry has recently come under fire for exorbitant pricing, even for drugs that have been on the market for a long time, such as insulin or epinephrine (e.g., EpiPens). In a recent analysis of the ten most expensive drugs in the US, the annual cost per drug ranged from $695,970 to over $2 million[47]. In addition, medical procedures and hospitalizations generate a great deal of healthcare costs. The average hospital stay costs over $10K. Procedures such as angioplasty can cost over $28K. Three of the most expensive operations, spinal fusion, heart bypass, and heart valve replacement, cost an average of $110K, $123K, and $170K, respectively. In 2021, the United States spent about $4.3 trillion on healthcare, an ever-increasing proportion of the gross domestic product. We are projected to spend $6.2 trillion by 2028, amounting to $18K per person and about 20% of the GDP. Most healthcare spending is intended to benefit the individual but not necessarily the genome; it may do just the opposite. Increasing numbers of genetically impaired individuals will likely exacerbate this situation, possibly within a few generations. How will societies afford to make effective healthcare available for everyone?

Ultimately, it is unclear whether the quality of life and longevity enjoyed by developed countries can be maintained. We live in a time of increasing global challenges: political polarization, severe natural disasters related to climate change, increased risk of pandemics, war, and as a result, increased social unrest and conflict as portions of the planet become uninhabitable. None of this bodes well for our ability to sustain the cultural and technological buffers that have protected us from natural selection in recent history. In the setting of a global catastrophe or unchecked climate change, the loss of the protective environments and medical support that we have crafted for ourselves and upon which we have become dependent could very quickly lead to massive suffering, illness, starvation, and death. In this scenario, natural selection could reassert itself. Recall that natural selection is not much influenced by human happiness or suffering unless they impact one's ability to survive and reproduce. Through premature death and failure to reproduce, natural selection would act to eliminate the accumulated load of mutations that render some of us less hardy, adaptable, and able to sustain ourselves. Even without the large-scale failure of society, it's not difficult to imagine the resurgence of selection for disease resistance should we begin to face large outbreaks of antibiotic-resistant bacteria, experience pandemics such as Covid-19, or fail to effectively vaccinate the population. Some authors have argued the increased genetic variability that characterizes the contemporary human genome can be beneficial from the standpoint of the species in allowing for more adaptive and evolutionary potential. Perhaps so, but evolution in response to renewed selective pressures would still involve differential survival and reproduction, with the associated death and suffering as our genome is pruned and re-shaped to favor those individuals who are genetically most fit.

In the following chapters, I will summarize what we know about the recent evolutionary history, genetics, and future concerns for some specific human traits. This illustrates the genetic and demographic complexities, the ongoing effects of natural selection or the lack thereof, and the environmental factors that influence current human evolution.

Sources:

Budnik, Alicja and Maciej Henneberg 2017. Worldwide increase of obesity Is related to the reduced opportunity for natural selection. PLoS ONE 12(1): e0170098. https://doi.org/10.1371/journal.pone.0170098

Henneberg, M. 1976. Reproductive possibilities and estimations of the biological dynamics of earlier human populations. Journal of Human Evolution 5: 41–48.

Loos, Ruth J. F. and Giles S. H. Yeo 2022. The genetics of obesity: from discovery to biology. Nature Reviews Genetics 23: 120-133.

Lynch, M. 2016. Mutation and human exceptionalism: our future genetic load. Genetics 202: 869–75.

Post, R. H. 1971. Possible cases of relaxed selection in civilized populations. Humangenetik 13: 253-284.

Roser, Max 2020. Our history is a battle against the microbes: we lost terribly before science, public health, and vaccines allowed us to protect ourselves. Our World in Data, https://ourworldindata.org/microbes-battle-science-vaccines. Accessed February 1, 2023.

Ross, Cody T. and Peter J. Richerson 2014. New frontiers in the study of human cultural and genetic evolution. Current Opinion in Genetics & Development 29: 103–109.

Ruhli, F. and M. Henneberg 2001. Biological future of humankind – ongoing evolution and the impact of recognition of human biological variation. In: Tibayrenc, M. and F. J. Ayala, eds. On Human Nature Biology, Psychology, Ethics, Politics, and Religion. London: Elsevier.

Stephan, C. N. and M. Henneberg 2001. Medicine may be reducing the human capacity to survive. Medical Hypotheses 57(5): 633-637.

Templeton, Alan R. 2016. The future of human evolution. In: Jonathan B. Losos and Richard E. Lenski, eds. How Evolution Shapes Our Lives: Essays on Biology and Society. Princeton University Press.

CHAPTER 9: Implications For Fertility

Fertility is a complex trait with significant genetic variance and evidence for rapid recent evolution[48]. From the standpoint of fitness, lifetime reproductive success depends upon both the number of children and the timing of reproduction. All else being equal, having more viable offspring and having them earlier in life will be associated with greater fitness. In a twin study from a contemporary Western population, early age at first reproduction and later age at menopause were found to be correlated with higher fitness. In contrast, age at menarche demonstrated an optimum, with fitness falling off at both younger and older ages. Recent evidence suggests that persons with a reproductive advantage associated with genetic predisposition for earlier age at first birth continue to be favored by ongoing natural selection in contemporary populations. However, the evidence for strong purifying selection around an optimum age suggests that the fitness advantage of early reproduction falls off at excessively young ages.

In recent years, it has become apparent that human reproductive behaviors are at odds with expectations based on historical selection for reproductive success. Although the issues that currently influence human reproduction are complex, one thing is clear; in addition to underlying genetic and biological potential, contemporary human reproduction has been dramatically impacted by both greater opportunities for individual reproductive choice and changes in the social/cultural environment. Progress in medicine and public health over the past 200 years has unintentionally allowed for human choice to override the evolutionary pressures that historically defined reproductive success. For most species, increased nutrition, health, and survival should lead to increased realization of genetic and biological potential resulting in greater numbers of offspring and consequent population growth. For humans, however, the benefits of cultural evolution have had the opposite effect during what has been termed the "demographic transition" of industrialized societies, paradoxically reducing individual fitness. Before the demographic transition:

> "...life was short, births were many, growth was slow, and the population was young. During the transition, first mortality and then fertility declined, causing population growth rates first to accelerate and then to slow again, moving toward low fertility, long life, and an old population. The transition began around 1800 with declining mortality in Europe. It has now spread to all parts of the world and is projected to be completed by 2100...[when the number of]...births per woman will have dropped from six to two. ...Age at first marriage and first birth are generally moving to older ages throughout the industrial and much of the developing world[49]."

The onset of decreased mortality coincided with advances in medicine, public health, and preventive care. However, the causes of the onset of reduced fertility have been the subject of much discussion. One thing is clear: the underlying causes of contemporary reduced fertility derive from cultural, non-genetic influences. Unlike genetic inheritance, cultural transmission can be profound and rapid enough to account for the magnitude and speed of changes occurring during the demographic transition. In addition, human traits such as level of education, wealth, social status, and intelligence have complex and changing relationships with reproduction over time. In some instances, they may have opposite effects on males and females. Overall, the availability of inexpensive and efficient birth control methods has had one of the most profound effects on these relationships. When wealth is combined with women's greater empowerment and ability to control their own fertility, they often opt for a higher quality of life with opportunities for education, career, increased leisure, and personal comforts over reproduction.

One of the most profound choices related to reduced fertility is the unprecedented postponement of childbearing, reflecting the tradeoff between reproduction at the biologically optimal time versus delay so women can obtain additional education necessary for building a career, as well as a host of other reasons. However, this delay has a direct negative effect on reproductive fitness since other individuals who may reproduce earlier and more often will contribute more to future generations. In addition, when reproduction is delayed to a later age, women's fecundity may be in decline, and complications of gestation and delivery are more frequent.

Family planning programs that offer education and easy access to birth control have a major impact:

> "One large study that began in 1977 divided 141 villages in the Matlab region of Bangladesh into two groups. In half of the villages, female reproductive-health workers went to the homes of married women of childbearing age and offered free contraceptives and other health services. Women in the other villages were not visited and had to make the effort to go to government clinics to ask for contraceptives...In the villages visited by health workers, the fertility rate fell by 25% in the first two years of the programme, an effect that lasted for at least 20 years. These women had, on average, 1.5 children fewer than women who had to visit a clinic to get contraceptives. Rates of maternal mortality fell by 50% in villages visited by health workers, and women in these villages were less likely to be underweight than women who were not given easy access to contraception[50]."

One hundred fifteen countries had family planning programs in place by the 1990s. Various studies have shown that these programs and/or easy access to contraception are associated with delayed marriage and onset of reproduction, a greater likelihood of college education and employment for women, and higher wages for women. There are also benefits for their children: they experience better health with lower mortality, get more schooling, and are less likely to experience poverty in adulthood.

Denmark is an example of a country with a high standard of living but a low birth rate, below replacement at an average of 1.7 births per couple. One out of five men may be expected to produce no offspring. People appear to be largely content with their lives due to the rewards and satisfactions from other pursuits such as careers, hobbies, and travel, which might have to be traded against the investment required to raise children. One economist has also linked this trend to declining religiosity, leading people to seek external validation through work, thereby increasing its worth relative to reproduction. A recent study has projected that the global total fertility rate will be well below replacement levels at 1.66 children per female in the year 2100. Because of declining fertility, the global population is projected to peak at 9.73 billion people in 2064 and decline to 8.79 billion in 2100. Accompanying these total numbers would be a marked shift in age structure, with 2.37 billion individuals over age 65 and only 1.70 billion below age 20.

Ultimately, we are choosing to be less fit! This is not to imply, however, that the evolution of fertility no longer occurs in humans. Individuals with greater fertility will still be favored since they will contribute more to future generations. But in post-transition societies, differential fertility is largely determined more by behavioral and cultural influences rather than genetic variability in the biological potential for reproduction. The total number of births has been found to correlate between generations, so individuals with many siblings tend to have many offspring, largely for cultural/behavioral reasons, and their progeny will predominate in the population.

Because of both genetic and social transmission of fertility, it has been estimated that the majority of people alive today may be derived from less than 5% of people from ten generations earlier. Some clear-cut examples of the impact of social transmission involve the influence of religion and personal beliefs. In the United States, Orthodox and Hasidic Jews, Mormons, and Mennonites have higher birth rates than average. Based on fertility, Roman Catholic women have about 20% higher fitness than women in other religions. Among Protestants in the United States, conservative women have higher birth rates and earlier first births than others. Over decades, this greater fertility has accounted for over three-fourths of the observed growth in conservative congregations as compared with other Protestant denominational affiliations.

What are the implications for our genome? Differential fertility can still lead to differential fitness in an evolutionary sense. Yet considerable genetic variability has been found to underly life-history traits that ought to be under strong selection to optimize reproduction. The genotypes of those individuals who reproduce more will still tend to predominate in future generations, but it appears that selection favoring those individuals is more likely to be based on behaviors "inherited" largely by cultural transmission rather than being based primarily on genetics. In post-transition societies, genes that influence reproductive potential will be less likely to be impacted by either positive or negative natural selection since reproductive choice, rather than biological potential, has a greater effect on the number of progeny and therefore, fitness. Might this partly account for the degree of genetic variability observed in reproductive parameters such as age at menarche, age at first reproduction, and age at menopause? In the setting of relaxed selection, this genetic variability may reflect a gradual decrement in our average reproductive potential over time. At the same time, qualities such as higher education, wealth, social status, and intelligence (associated with reduced fertility due to their effects on the timing and amount of reproduction) may be at an evolutionary disadvantage[51].

Sources:

Kirk, Katherine M. et al. 2001. Natural selection and quantitative genetics of life-history traits in western women: a twin study. Evolution 55(2): 423–435.

Sohn, Emily 2020. Planning for success. Nature 588: S162-S164.

Sussman, Anna Louie 2019. The End of Babies. The New York Times, November 16, 2019.

Vollset, Stein Emil et al. 2020. Fertility, mortality, migration, and population scenarios for 195 countries and territories from 2017 to 2100: a forecasting analysis for the Global Burden of Disease Study. The Lancet 396(10258): 1285-1306.

CHAPTER 10: Implications For Infertility

Epidemiology

The medical treatment of infertility[52] is another aspect of human reproduction that may have evolutionary consequences more directly related to the circumvention of natural selection. It is surprisingly common, affecting around 17.5% or one out of six couples globally. In women, the causes include ovulation disorders, endometriosis, early menopause, and anatomical abnormalities such as fallopian tube damage resulting from sexually transmitted infections. For men, the causes include abnormal sperm production or delivery and environmental exposures such as pesticides, alcohol intake, or excessive heat. For both sexes, genetic abnormalities underly a significant portion of cases. Based on a survey conducted in the United States from 2006 to 2010, infertility affected 6%, or 1.5 million women aged 15 to 44. An additional 11% of women had impaired fecundity, defined as physical difficulty in either getting pregnant or carrying a pregnancy to live birth. The Centers for Disease Control reports that one in five women with no prior births have infertility issues[53]. Globally, about 2-10% of women have fertility problems with either their first or subsequent pregnancies. A survey reported in 2007 estimated that 72.4 million women worldwide were experiencing infertility, with 40.5 million actively seeking medical care.

As suggested by these epidemiological data, infertility is often considered primarily a female issue. In reality, male and female causes each account for about one-third of cases, and the remaining third may involve both sexes, or the cause may be unknown. The rates of male infertility are in the range of 2.5-12% and increasing. Several other impairments of male reproductive health, such as testicular germ cell cancer, undescended testes, and penile hypospadias (abnormal urethral opening on the underside of the penis rather than at the tip), also appear to be on the increase. Both genetic and environmental/lifestyle causes have been implicated. It is noteworthy that male infertility is associated with other medical disorders and reduced life expectancy. For example, a comparison of over 36,000 healthy and infertile men from a national insurance database in the U.S. reported in 2016 that men diagnosed with infertility also had a greater risk of subsequently developing diabetes, coronary artery disease, alcohol abuse, and drug misuse disorder.

The medical management of infertility has grown with increased demand, public awareness, and advances in medical technology. Women seeking medical assistance for fertility problems are more likely to be married, non-Hispanic white, older, highly educated, and more affluent than nonusers of fertility services. During 1982–2010, 12% of women in the U.S. used infertility services of some kind, including advice, infertility tests, drugs to regulate ovulation, or medical assistance in preventing miscarriage. Nearly 9% of women used medical assistance to become pregnant.

The latter category is known as assisted reproductive technology (ART) and involves hormonal treatments, in-vitro fertilization (IVF) and/or intracytoplasmic sperm injection (ICSI). For both IVF and ICSI, fertilization takes place outside of the body in a laboratory setting. The number of ART-related procedures performed in the US tripled between 1996 and 2015. In 2015, 182,111 ART procedures resulted in 71,152 infants born. The rate was 2,832 procedures per 1 million women of reproductive age, contributing to 1.7% of all infants born. In Denmark, 8% of all children are born after ART, contributing significantly to the total number of births. It is currently estimated that more than 5 million babies worldwide are the products of ART, and in many Western countries, access to infertility treatment is considered to be a universal right.

Genetics Of Reproduction

Our primary concern in this discussion is the possibility of genetic consequences from treating human infertility. We will begin by looking at what is known about the genetics of reproduction and the processes involved are complex. In females, successful reproduction requires anatomical and functional development of female reproductive organs, including the ovaries, development and maturation of eggs, successful fertilization, early development of the fertilized egg, successful implantation, and fetal growth. These processes depend upon precise hormonal interactions among the hypothalamus, pituitary, and reproductive tract, as well as the ability to carry a pregnancy to term. Overall, it is likely that 200 or fewer genes account for most cases of female infertility. Multiple genes are involved in the production and maturation of eggs, and primary ovarian insufficiency, the failure to produce viable eggs, occurs in about 1-2% of women. In addition, genetic abnormalities of the X chromosome that occur in 5-6% of women can cause recurrent fetal loss.

Nearly 30 million males worldwide have infertility, although there is wide variation in the prevalence among different geographic regions. Infertility can be associated with undescended testes, varicocele (abnormal enlargement of the venous plexus in the scrotum), hormonal disorders, obstruction/absence of seminal pathways, infections, alcohol consumption, or chemotherapy. Familial patterns of male infertility have long been noted, and it is estimated that up to half of all cases are associated with genetic defects. Two thousand or more genes may be involved in the production of sperm, of which more than 500 may be associated with infertility. They can affect the development of the testes or urogenital tract, the production and maturation of germ cells (which produce sperm), or the functionality of sperm cells. Chromosomal abnormalities and microdeletions affecting a portion of the Y chromosome, called the long arm, are the leading causes of male infertility. The Y chromosome is particularly interesting since it contains genes responsible for testis development and the initiation and maintenance of sperm production in adulthood. The long-arm portion seems to be predisposed to harmful deletions; abnormalities in a specific portion of the long arm called the Azoospermia Factor (AZF) region may be associated with 7.5% of cases of male infertility.

The genetics of the AZF region have been relatively well studied. This region has been functionally subdivided into three distinct, non-overlapping sub-regions, which are designated AZFa, AZFb, and AZFc. Abnormalities affecting these sub-regions are one of the leading causes of the failure of sperm production. Many males with extreme testicular pathologies and/or severely reduced sperm numbers carry Y chromosome microdeletions, with AZFc being the most frequently deleted portion. For example, one deletion designated gr/gr occurs in 5-15% of men and is associated with low sperm counts. However, its prevalence, as well as its association with infertility, varies among populations. In one study, infertile men were found to be twice as likely to carry gr/gr. In addition to infertility, DNA damage in sperm can be associated with pregnancy complications. It is also negatively correlated with the success of ART, although the effect can be partially overcome by using ICSI.

Not all genes affecting male fertility occur in the Y chromosome. At least 11 specific genes related to infertility have been identified on other chromosomes. For example, one gene designated SYCP3 on chromosome 12 codes for a protein necessary for a type of cell division involved in the production of sperm, and abnormalities in this gene result in the complete absence of sperm. There also appear to be many other DNA variants affecting fertility that are rare and/or yet to be discovered. Ten to fifteen percent of men at prime reproductive age suffer infertility, where the specific genetic cause remains unknown. Overall, mutations in several hundred different human genes could be involved, but each of them is likely to be responsible in only a very small percentage of cases.

Unintended Outcomes Of Art

The failure to reproduce directly reduces fitness, so there should be strong selective pressure against reproductively harmful genetic variants. Under normal circumstances, problems during conception and early pregnancy may result in a failed pregnancy up to 30% of the time. Often, this occurs even before a pregnancy is apparent and constitutes substantial early selective pressure favoring genetically viable fetuses. When medical interventions remove obstacles to successful reproduction, there is potential for natural selection to be circumvented and allow genotypes that are not optimal to proceed through pregnancy and possibly passed along to subsequent generations.

There is accumulating evidence that infants conceived by ART are at increased risk for significant complications and disorders. Risks during pregnancy include placenta previa, abruption, hemorrhage, hypertensive disorders of pregnancy, gestational diabetes, induction of labor, and Caesarean section. The delivery of infants conceived with ART is complicated by preterm delivery and low birth weight. Although this is associated with an increase in multiple gestations (e.g., twins, triplets) following ART, there remains as much as a two-fold increase in the risk of perinatal mortality, preterm birth, and low birth weight even among singletons or twins compared with those conceived naturally. Most concerning is a 30%–40% increase in the rate of major malformations for infants conceived through ART. Malformations include heart defects, brain/spinal cord defects, and digestive tract abnormalities. Although the reasons for these outcomes may be multifactorial, there is evidence that the increase in congenital malformations is related to parental characteristics (including genetics) rather than issues related to the ART procedure itself. Following birth, ART-conceived children may be at increased risk for being small for gestational age, developmental disability, impaired glucose metabolism and insulin resistance, elevated levels of thyroid-stimulating hormone, and cardiovascular diseases. If ART procedures facilitate a successful pregnancy involving sub-optimal genotypes, there is also a risk of disorders such as infertility being passed on to subsequent generations.

Among ART procedures, ICSI may pose the greatest risk for transmission of genetic abnormalities. With conventional IVF, fertilization occurs "naturally" outside of the body. In contrast, ICSI involves fertilization by injecting a single sperm directly into an egg, even though the selected sperm may have impaired mobility or abnormal morphology that would have prevented fertilization under normal conditions. As described by Ziru Jiang and colleagues at the Shanghai Jiao Tong University in China:

> *"ICSI...evades natural selection at the oocyte [egg] membrane, which occurs both during natural conception and in conventional IVF, and allows genetically and structurally abnormal sperm to fertilize eggs and pass abnormal genetic materials to the children[54]."*

ICSI has been associated with a greater incidence of undescended testicles, chromosomal abnormalities, autism, intellectual disabilities, hormonal abnormalities, and congenital disabilities. Nevertheless, the use of ICSI as part of IVF increased from 36.4% in 1996 to 76.2% in 2012. There are also clearly documented cases of the transmission of infertility to offspring. For example, one study examined 18 children conceived via ICSI whose fathers had fertility problems stemming from an AZFc deletion[55]. Ten of the eighteen offspring who were males had inherited this mutation. Similar risks include transmission of a genetic defect such as a cystic fibrosis mutation that causes male infertility, defective sperm activating factors, and potentially additional mutations affecting non-reproductive organs. Of particular concern is the possibility of transmission of the mutation responsible for bilateral absence of the vas deferens (the ducts which convey sperm to the urethra), which is one of the major indications for ICSI. There is also potential for transmission of genetic abnormalities associated with defective DNA repair.

A hypothesis proposed by Andrew Czeizel and Kenneth Rothman from the Boston University School of Public Health highlights one possible consequence of relaxed selection pressures relating to infertility[56]. They cited evidence that sperm density in the human population declined during the 20th century, particularly in Europe. Sperm count declined in Western countries at a rate of 1.4% per year between 1973 and 2011, with no evidence of leveling off. Environmental exposures had previously been invoked as a cause of these trends, but Czeizel and Rothman proposed that they could be a consequence of relaxed selection pressure. Citing population data from Hungary, they noted that the number of offspring per fertile couple declined from 11 to

1.5 during the 20th century, related to the demographic transition. At the same time, the ability of previously infertile couples to reproduce was enhanced due to the availability of new fertility treatments. Over time, the number of offspring from previously infertile couples tended to converge with the number from fertile couples. As a result, the proportion of children in the general population born to sub-fertile couples increased from 1-4% to about 14-20% by the end of the 20th century. They hypothesize that the relaxation of natural selection against subfertility due to medical advances may account for the decline in sperm density, as the proportion of men with an inherited tendency toward subfertility in the population has increased. Hypothetically, if half of all infertile males underwent successful ICSI, the incidence of severe male infertility could double in seven generations.

Although many healthy children are born as a result of treatment for infertility, it is clear that children conceived through ART are at greater risk for problems than children conceived naturally. The causative factors are not entirely understood, and there are likely contributions from the direct effects of the procedures themselves upon the gametes and early embryonic stages. However, there is also concern that medical interventions allow for the inter-generational transmission of parental or acquired mutations that otherwise would have been negatively selected via the mechanism of infertility. In addition, the combination of ARF screening processes that may favor individuals with certain socioeconomic characteristics, and a markedly altered environment for reproduction, creates a modified set of (natural) selection criteria that may have adverse long-term genetic consequences. The Norwegian gynecologist Hans Ivar Hanevik and his colleagues have summarized:

> *"Although IVF is a great medical achievement, it circumvents a range of pre- and post-zygotic reproductive barriers. It increases the reproductive fitness of subfertile couples by technologically removing several naturally occurring selective barriers and by altering other such barriers. In accordance with the basic principle of evolution, the subsequent generations will thus be genetically and epigenetically adapted to an environment in which reproduction is increasingly dependent on technological intervention. It is our opinion that IVF should be seen as a primary example of how the human species is becoming not only culturally —but also biologically—dependent on our own technology[57]."*

In a broader sense, the inheritance of infertility is also impacted by medical interventions during pregnancy and delivery. An example concerns the implications of Caesarian section for cephalopelvic disproportion (CPD) - which is responsible for most cases of obstructed labor. CPD occurs when there is a mismatch between the size of an infant's head (too large) and the size of the mother's birth canal (too small), such that vaginal delivery becomes dangerous or impossible for the mother and child. Humans may be particularly susceptible to this condition due to a historical tradeoff in selection pressures favoring large neonatal size, which is associated with improved infant survival, and a narrower female pelvic cavity associated with walking on two legs. In the absence of medical care, CPD often leads to maternal and neonatal death/disability, imposing selection against any size mismatch. The availability of a relatively safe alternative means of delivery (Caesarian section) provides for the survival of the mother and the infant, potentially carrying a genetic predisposition for similar problems into the next generation. One model of this phenomenon predicts a 10-20% increase in the rate of CPD over the course of two generations.

Medical interventions for infertility can potentially allow for the inter-generational transmission of inherited or acquired mutations that otherwise would have been subject to negative selection via the mechanism of infertility. Partially in recognition of these risks, genetic screening for at least one type of mutation in the AZF region of the Y chromosome is often included in the routine diagnostic testing of infertile men. Although this does not preclude the possibility of passing along genetic traits associated with infertility to subsequent generations, it does provide prospective parents with information regarding the risks of doing so and the opportunity to choose not to perpetuate such risks. Unfortunately, limitations in our current understanding of all the other genetic risk factors involved in infertility leave us with no such opportunities in most cases.

Sources:

Agarwal, Ashok et al. 2015. A unique view on male infertility around the globe. Reproductive Biology and Endocrinology 13: 37. https://doi.org/10.1186/s12958-015-0032-1.

Boivin, Jacky et al. 2007. International estimates of infertility prevalence and treatment-seeking: potential need and demand for infertility medical care. Human Reproduction 22(6): 1506–1512.

Boulet, Sheree L. et al. 2015. Trends in use of and reproductive outcomes associated with intracytoplasmic sperm injection. Journal of the American Medical Association 313(3): 255-263.

Chandra, Anjani et al. 2014. Infertility Service Use in the United States: Data From the National Survey of Family Growth, 1982–2010. National Institutes of Health, National Health Statistics Reports 73: 1-21.

Hansen, Michele et al. 2013. Assisted reproductive technology and birth defects: a systematic review and meta-analysis. Human Reproduction Update 19(4): 330–353.

Howard, Jacqueline 2023. Infertility affects a 'staggering' 1 in 6 people worldwide, WHO says. CNN Health, April 3, 2023. https://www.cnn.com/2023/04/03/health/infertility-global-prevalence-who-report/index.html.

Hwang, Kathleen et al. 2010. Mendelian genetics of male infertility. Annals of the New York Academy of Sciences 1214: E1–E17.

Levine, Hagai et al. 2017. Temporal trends in sperm

count: a systematic review and meta-regression analysis. Human Reproduction Update 23(6): 646–659.

McDonald, Sarah D. et al. 2005. Perinatal outcomes of singleton pregnancies achieved by In vitro fertilization: A systematic review and meta-analysis. Journal of Obstetrics and Gynaecology Canada 27(5): 449-459.

Miyamoto, Toshinobu et al. 2017. Human male infertility and its genetic causes. Reproductive Medicine and Biology 16: 81–88.

Qin, Jia-Bi et al. 2017. Worldwide prevalence of adverse pregnancy outcomes associated with in vitro fertilization/intracytoplasmic sperm injection among multiple births: a systematic review and meta-analysis based on cohort studies. Archives of Gynecology and Obstetrics 295: 577–597.

Swan, Shanna H. 2021. Reproductive Problems in Both Men and Women Are Rising at an Alarming Rate. Scientific American Newsletter, March 16, 2021. https://www.scientificamerican.com/article/reproductive-problems-in-both-men-and-women-are-rising-at-an-alarming-rate/

Yatsenko, Svetlana A. and Aleksandar Rajkovic 2019. Genetics of human female infertility. Biology of Reproduction 101(3): 549–566.

CHAPTER 11: Implications For The Immune System

It is generally accepted that infectious diseases have imposed profound selective pressures during human history, particularly since the transition from hunter-gathering to farming beginning about 10,000 years ago. The clustering of human populations into larger groups facilitated the spread of infections among us, as well as exposure to human diseases derived from domesticated animals or pests such as rodents. The situation would have been further complicated by problems associated with mass waste accumulation and potentially unreliable access to safe water supplies. The most evolutionarily significant infectious diseases include:

cholera	tropical yellow fever
dengue hemorrhagic fever	hepatitis B
East and West African sleeping sicknesses	influenza A
malaria	measles
visceral leishmaniasis	pertussis
diphtheria	rotavirus A
mumps	syphilis
Plague (Black Death)	tetanus

rubella	tuberculosis
smallpox	Chagas' disease
typhoid and typhus	

Demonstrations of the magnitude of selection imposed by these diseases abound. The Black Death has been described as the single greatest mortality event in recorded history[58]. The first outbreak of the second pandemic of this plague killed about half of Europe's population during the 14th century. The introduction of old-world pathogens during European colonization of the new world devastated poorly adapted indigenous populations with no prior exposure. For example, the Aztec empire suffered a severe salmonella epidemic that was introduced during the Spanish conquest in the early 1500s, and as much as 80% of the population was lost. Indigenous populations in North America and Greenland numbered in the millions prior to the year 1500 but were reduced to around 375,000 by 1900 due to disease and other impacts of colonialism. The Spanish flu pandemic of 1918-1920 is believed to have resulted in 20 million or more deaths worldwide. Smallpox is estimated to have killed at least 300 million people in the 1900s alone, leaving many survivors scarred and blind. Up to the mid-19th century, infectious diseases killed half of all children by the age of 15, limiting average life expectancy to about 25 years. Although now reduced, infectious disease continues to have a significant impact, accounting for about 15% of deaths worldwide and up to 41% in parts of Africa.

In response to intense selective pressure, vertebrate adaptation has resulted in a complex system to combat invading pathogens. The first line of defense is the innate immune system. It consists of several general mechanisms to prevent or respond to infection. The first line of defense consists of barriers that may be anatomical, such as skin; chemical, such as acids within the digestive system; or biological, such as the cell layers forming the barrier between the blood and the brain. Inflammation is one of the earliest responses to invasion of a pathogen. Specialized cells called macrophages can detect pathogens and release chemical mediators that initiate the inflammatory response: dilation of blood vessels, sensitization of pain receptors, and attraction/activation of white blood cells such as neutrophils that can combat pathogens. The ability to clear pathogens is also facilitated by the complement system – a chemical cascade that recruits more inflammatory cells, identifies pathogens, and damages/clears pathogen cells.

These defenses are supplemented by the adaptive immune system. It is composed of specialized cells and processes that help eliminate specific pathogens and retain a "memory" that allows for a faster and stronger immune response upon re-exposure to the same pathogen. Antigens are molecules usually associated with a pathogen and are recognized by the immune system as "non-self," triggering the immune response. Within the adaptive immune system, specialized white blood cells called B cells recognize antigens and become activated. They divide and produce offspring that secrete antibodies that bind to the specific antigens attached to pathogenic cells, marking them for destruction and removal. Other specialized cells called killer T cells identify and kill cells that have become infected with pathogens. Overall, the adaptive immune system is an effective means to combat specific pathogens and prevent reinfection by the same pathogens in the future. It also provides the immune basis for vaccination - intentional injections of dead or weakened pathogens or their antigens to activate a fast and effective immune memory response in the event of re-exposure to the same pathogen.

The genetics underlying all these immune defenses is complex and extensive. Most selective events affecting genes that are involved in innate immunity occurred during the interval 6,000-13,000 years ago, consistent with the human transition to agriculture, although some events may have occurred as recently as 2,600-1,200 years ago. Genes with evidence for positive selection are often found to have regulatory effects, altering the magnitude of gene expression and responses to immune stimuli.

The results of a recent analysis of ancient DNA extracts from the time of the Black Death provide evidence for a recent and powerful selective effect of infectious agents[60]. Comparing immune-related genes from individuals who died shortly before, during, or soon after the Black Death (which occurred from 1346–1350) in London and across Denmark, four genetic variants were identified that appeared to have undergone strong positive selection resulting from the pandemic. One of the variants, designated ERAP2, affects the chemical response to *Yersinia pestis* and increases the ability of macrophages to fight the pathogen. Those individuals who carried two copies of the protective variant are estimated to have had 40% greater survival than those who carried no copies[61].

Variation among individuals in response to a

pathogen is extreme: exposure to the same pathogen in different individuals may cause no symptoms, produce mild to moderate illness, or be lethal. There are many possible reasons for such diversity of responses, but underlying variation in genetics is believed to account for much of it. For example, in some cases, specific genetic defects known as primary immune deficiencies cause susceptibility to specific diseases in otherwise healthy individuals. This single-gene type of immune deficiency tends to primarily affect the risk in children. We currently know of more than 350 genes involved with such defects, and the number is increasing with time.

Susceptibility to disease in adults is more likely to be polygenic. Genome-wide association studies have revealed at least 200 immunity-related genes with extensive variation resulting from adaptation against a wide variety of pathogens. The major histocompatibility complex (MHC) is a region of immune genes located on chromosome 6 that code for proteins involved in the recognition of antigens, leading to the destruction of infected cells, activation of macrophages, and production of antibodies. MHC genes manifest the greatest degree of genetic variability known in the human genome. Incredibly, more than 2700 variants have been documented for a single gene in this region. Genetic analyses have determined that the creation and maintenance of this genetic diversity has included both positive and balancing selection for enhanced resistance to pathogens and the introduction of beneficial variants through ancient interbreeding with our hominin cousins.

The degree of genetic variation within the human immune system is so extensive that it must be associated with a wide degree of variability in the susceptibility to different infectious diseases. Hypothetically, one may conceive of an ideal combination of genetics to protect from most or all infections. From this perspective, it has been suggested that every time someone contracts an infectious disease, it may be partly related to some relative, genetically based weakness in our immune system.

Considering the evolutionary origins of our immune-related genes reveals a paradox: even though there is ample evidence for the impact of positive selection, we appear to possess many harmful variants associated with common human immune diseases. Part of the explanation may involve positive selection for genetic diversity itself to combat a wide variety of pathogens. Within such diversity, there may well be variants that are maintained at significant frequencies even though they may be mildly harmful. This is related to a concept we encountered previously called antagonistic pleiotropy: where genetic variants have opposite effects on different traits. It is believed that the need to combat pathogens in human history counterbalanced the potentially harmful effects of variants that also increased the risk of immune-related diseases so that those variants were positively selected because of the overriding benefits of an aggressive immune system for protection against infectious challenges. For example, there is evidence for recent positive selection in seven different immune genes also associated with systemic lupus erythematosus. In one study, all the genes associated with resistance to *Mycobacterium leprae* (the cause of leprosy) are also implicated in inflammatory bowel disease. Two genetic variants believed to be favored by conferring resistance to Protozoan pathogens have each been found to affect susceptibility to 6-7 immune-related diseases. Other disorders associated with immune genes subject to positive selection in the past include asthma, celiac disease, multiple sclerosis, rheumatoid arthritis, ankylosing spondylitis, psoriasis, and type 1 diabetes. Clearly, there is inherent disease risk in our collection of immune genes.

In contrast to most of our evolutionary history, contemporary advanced societies are no longer subject to intense levels of natural selection associated with the morbidity and mortality from infectious disease[62]. Increased life expectancy over the past one or two centuries primarily resulted from the control of infectious diseases based on the increased availability of adequate nutrition, sanitation, safe drinking water, vaccines, anti-infectious drugs, and improved supportive care. In Italy, for example, the incidence of all contagious and parasitic diseases dropped from 136.2 per 100,000 inhabitants in the late 1940s to only 3.6 per 100,000 in 1990. Epidemics of polio and measles that were rampant 100 years ago are now nearly unknown in developed countries, and we now have effective vaccines against at least 30 diseases. This relaxation of natural selection is likely to increase genetic variability, including increased variability in susceptibility to disease. The elimination of the previous selective benefits of variants with a history of both beneficial and harmful immune effects leaves harmful effects as their primary manifestation. One might expect their frequency in the population to decrease over time due to negative selection. But, in addition to improving medical approaches to avoid or treat infections, we have also improved our management of immune disorders. To the extent that afflicted individuals now survive and reproduce, the expected negative selection against variants that cause immune disease is weak or absent. Other variants that may increase susceptibility to diseases such as polio, smallpox, mumps, or measles are also likely to have little impact on fitness since morbidity and mortality from these infections are either negligible or absent, and those variants may now be more likely to be transmitted from generation to generation. The coevolutionary relationship between humans and their pathogens can amplify the problem. As expressed by the eminent population geneticist James F. Crow in 1968:

> *While the problem of genetic load* [i.e., load of harmful variants] *may not be immediate because of its slow accumulation, the problem becomes more apparent for infectious disease. Medical treatment not only enables less favorable genes to accumulate in the human population over time, but also exerts a selective pressure on disease processes/pathogens...Consequently, humans may be becoming less fit faster, since unfavorable genes will accumulate in the human gene pool while pathogens become more genetically fit and more virulent*[63].

From the standpoint of public health, the disease risk is not trivial. Dysregulation of the immune system already poses significant health challenges in our society. It can lead to an excessive reaction in response to an immune trigger such as an allergen, or at times even in the absence of a trigger. Autoinflammatory diseases occur when unprovoked activation of the innate immune system produces symptoms that mimic active infection. One example is Familial Mediterranean Fever, which is associated with recurrent attacks of fever, abdominal pain, chest pain, and joint pain. In other cases, dysregulation can lead to attacks directed at the wrong target. Autoimmune disease results from problems within the adaptive immune system. It is essential that the specialized cells of this system be able to distinguish foreign antigens, such as those presented by pathogens, from the profiles of normal human body tissues ("self"). If the immune system mistakenly identifies normal tissues as foreign antigens, these cells may activate to attack the normal tissues and produce antibodies against them. In type 1 diabetes, for example, the adaptive immune system attacks the insulin-secreting cells of the pancreas. In systemic lupus erythematosus (SLE), failure to recognize DNA and its associated proteins as "self" leads to widespread damage. If the reaction is directed against harmless bacteria in the gut, inflammatory bowel diseases such as ulcerative colitis or Crohn's disease may result. In many cases, the initial trigger may be infection by a virus, bacterium, or parasite, resulting not only in the development of antibodies to the pathogen but cross-reaction with normal body tissues leading to an autoimmune response. The genetic basis is generally polygenic. For example, genome-wide association studies have documented 200 risk variants for inflammatory bowel disease and 101 risk variants for rheumatoid arthritis. Many of these variants also affect the risk of more than one immune disease.

The incidence of autoinflammatory and autoimmune

disorders is increasing. Worldwide, they affect 7.6 to 9.4% of the population and are among the leading causes of death for young and middle-aged women in the US and the UK, with rates of increase in the range of 10 to 20% a year over the last 30 years. Childhood food allergies in the U.S. have increased by 50% during the decade ending around 2010. During the same period, skin allergies increased by 69%. Childhood asthma increased by 38% between 1980 and 2003. In addition, we are seeing significant worldwide increases in the rates of specific autoimmune disorders such as inflammatory bowel diseases, type 1 diabetes, and multiple sclerosis. The trend toward increased incidence of diseases resulting from immune dysregulation is most consistently found in industrialized countries with temperate climates, where infectious diseases have generally been decreasing over time. For example, food allergies in children occur at a prevalence of 10% in Western countries compared to 2% in China. New cases of type 1 diabetes occur from 10 to over 100 times more frequently in Finland than in countries such as Mexico or Pakistan. Ulcerative colitis is twice as common in Western Europe compared with Eastern Europe.

Pinning down the magnitude of genetic contribution to increasing autoinflammatory and autoimmune disorders is difficult since these disorders are likely to result from complex interactions between genes, alterations in immune regulation, and environmental factors. For example, many authorities have suggested that the increasing problem of inflammatory and autoimmune diseases has resulted from contemporary hygiene practices that have dramatically altered our environment[64]. The so-called "hygiene hypothesis" was first formulated in 1989 by the epidemiologist David Strachan of the London School of Hygiene and Tropical Medicine. He surveyed over 17,000 British children and noted that those with elder siblings, whose presence likely increased early exposure to potential pathogens, were less likely to develop eczema in the first year of life or hay fever later in life. Additional protective effects against allergic diseases have also been noted for low antibiotic consumption, early pet exposure (especially to dogs – our best friends in so many ways), daycare attendance, or growing up on a farm.

The hygiene hypothesis recognizes that humans in the past have had to exist in environments that were rife with pathogens. Abundant exposure to infectious and parasitic organisms, particularly during childhood, may have played an essential role in moderating our immune system since co-existence with these organisms meant that they had to be tolerated to some extent to avoid immune-mediated damage to ourselves. But recent economic development and a focus on reducing exposure to infectious diseases have significantly altered our environment. We mop our floors, sanitize kitchen and bathroom surfaces, wash our hands, pasteurize our milk, refrigerate food, monitor and control food production, quarantine, take antibiotics, vaccinate, avoid vectors of disease transmission such as mosquitoes, purify our water and isolate/treat human waste. In doing so, we have profoundly changed our microbial environment and reduced infections such as mumps, measles, tuberculosis, malaria, hepatitis A, smallpox, and parasitic worms. In our more "sterile" environments, the effects of early pathogen exposures, which over preceding millennia served to moderate our future immune responses, are absent, setting the stage for over-reaction of the immune system when it does encounter environmental or biological triggers. The increasing incidence of type 1 diabetes in industrialized countries may be one example of the adverse effect of declining infections on the activity of the immune system.

Graham Rook, an immunologist at the University College of London, has emphasized the coevolutionary relationships between humans and some potential pathogens[65]. Coevolution occurs when each organism adapts in response to the long-term presence of the other organism in its environment. Historically, we have had to tolerate the great diversity of our microbiota[66] without initiating a potentially damaging immune response. The presence of these organisms, called "old friends" by Dr. Rook, is thought to help "train" the immune system to identify and react appropriately to friends and foes. In contrast to the hygiene hypothesis, this mechanism of immune moderation does not require childhood infection. To optimize their fitness, each organism has had to adapt to the presence of the other. Because of this, some parasites have evolved to play an active role in moderating our immune system in order to preserve their ability to survive and reproduce. We have had to accept and tolerate these organisms within our bodies without activating a potentially self-damaging immune response. Overall, coevolutionary relationships may have come to play an essential role in promoting normal human immune development. Increased emphasis on hygiene can adversely affect our exposure not just to pathogens but to "old friends" as well. The result may be an increased likelihood of immune over-activation.

One of the best-characterized examples of

coevolutionary influences on our immune system involves small parasitic worms known as helminths. These include flukes, tapeworms, and roundworms that commonly infected humans throughout our history and, even today, may affect up to 1 to 2 billion individuals. As of 2008, it was estimated that helminths infected 37% of the entire world population, mainly in developing countries. Parasitic schistosome worms such as the blood fluke *Schistosoma mansoni* currently infect about 200 million people in tropical and subtropical regions. Humans become infected when they come into contact with immature larvae, which penetrate the skin. The larvae then migrate to the lungs and the liver, where they mature. After mating, the females produce thousands of eggs, which exit the body through the bladder or intestines. The eggs then hatch in fresh water, where the larvae infect snails - their secondary host. There they multiply through asexual reproduction (no males required!) and become available to repeat the cycle once again. Individual schistosomes can live within humans for up to 40 years. In areas where schistosomiasis is common, people can become infected in early childhood and may remain infected for their entire lives.

The remarkable long-term relationship between schistosomes and their human hosts has led to coevolution. Schistosomes have evolved to protect themselves against an aggressive host immune response through adaptations that moderate that response. They coat themselves with human antigen to avoid detection, and they can destroy human antibodies. They also secrete substances that suppress both the innate and adaptive arms of the human immune system. These adaptations are fine-tuned to avoid excessive suppression of the host response since the parasites benefit from having their host survive as long as possible. Humans, in turn, have evolved to combat helminth infections. Some 3478 human genetic variants have been identified that correlate geographically with helminth diversity. They are found in 810 distinct human genes, some of which have also been linked to increased susceptibility to autoimmune or inflammatory disorders such as asthma, allergy, celiac disease, and inflammatory bowel disease. The helminth-mediated immune suppression not only protects the parasites within their human host but also limits overall immune-related damage to the human host that could result from long-term immune activation.

One result of the evolutionary arms race between humans and their helminth parasites is an ongoing dynamic balance between the host's immune response and parasite-mediated immune regulation. To the extent that the prevention of immune damage to the host depends upon the checks and balances provided by the parasite, the absence of the parasite could become a significant liability. For example, studies in South America and Africa have found a negative correlation between the incidence of helminth infections and the presence of allergies or asthma. Uninfected patients have a five-fold higher proportion of positive allergic skin tests than infected patients. Given this, there is concern hygienic practices and medical treatment that have nearly eliminated helminth infections in developed countries result in an anti-helminthic immune response that is both unsuppressed and unopposed. In experimental animal models, infections with schistosomes have been found to prevent type 1 diabetes, an encephalomyelitis that is analogous to human multiple sclerosis, and Graves' disease, an autoimmune disorder affecting the thyroid gland. Patients with multiple sclerosis who are also parasite-infected exhibit a significantly lower number of flare-ups, more stable functionality, and fewer nervous system changes apparent on magnetic resonance imaging (MRI) compared with uninfected patients. It has even been suggested that the immune response triggered by helminth infection may be protective against cancers. No one suggests that we abandon practices of good hygiene, which help to protect us against harmful pathogens. Nevertheless, there has been interest in ingesting helminth eggs (*Trichuris suis*) as a treatment for Crohn's disease and ulcerative colitis. A more acceptable possibility could be to harness the secretory products produced by helminths and other pathogens that moderate our immune response.

There has been increasing interest and recognition of the role of our microbiome in regulating the development of our immune system. The microbiome is an abundant collection of mostly benign or beneficial microorganisms that colonize our internal and external surfaces, such as our skin, oral and vaginal mucosa, and the gastrointestinal tract – particularly the colon. The effects of this community of microorganisms have been incorporated into our development and physiology, particularly with regard to the immune system, due to likely extensive coevolution over time. We initially acquire the organisms that compose the microbiome from our mothers beginning in utero and later during passage through the birth canal and during breastfeeding. Beyond the birth process, children continue to accumulate microbes when they encounter family members, friends, pets, impure water, and dirt. Exposure to these microbial communities occurs during the same period of time that we undergo significant development of our immune system.

Common interventions employed in modern societies to improve our health and safety may have opposite and unintended long-term effects due to their impact on our microbiome during the development of the immune system. The microbial community can be significantly altered due to pre or postnatal antibiotic exposure, Caesarean delivery (where the microbial community colonizing the mother's birth canal is bypassed), or when the mother does not breastfeed. Cesarean delivery may occur in as many as one-third to one-half of births and has been associated with a higher risk of asthma, type 1 diabetes, multiple sclerosis, and celiac disease. In one study, early antibiotic exposure was found to be associated with an increased risk of childhood-onset asthma, allergic rhinitis, atopic dermatitis, and celiac disease. There were also associations with possible non-immune outcomes: obesity and attention deficit hyperactivity disorder. Another study has suggested that early antibiotic exposure increases the risk for the development of inflammatory bowel disease. Alterations in microbial composition caused by antibiotics have the potential to not only affect immune and other physiological development but also to alter the development and normal maturation of the adult microbiome. The overall effect tends to be an overly active immune system. Even more concerning is the possibility that antibiotic-induced alterations in our microbiome may be passed down over generations, so that loss of our microbial diversity could steadily worsen.

A comparative study of genetically similar populations from northern Europe provides evidence for the potential impact of the gut microbiome on immune function[67]. Stool samples of 222 children between the ages of one and three were used to characterize the gut microbiome's species. The results for children from Estonia and Finland – relatively modernized societies where early-onset autoimmune disorders are common – were compared with those in children from Russia – a relatively less modernized region with a much lower incidence of such disorders. Bacteria of the genus *Bacteroides*, which inhibit the immune system early in life, were found to be dominant in the samples from Estonia and Finland. In contrast, samples from Russia were primarily found to have *Escherichia coli*, a potent immune activator. These findings suggest that an early immune response to gut microbiota may be protective of later autoimmune disorders. In effect, the absence of pathogenic organisms present throughout our evolutionary history has compromised our immune system. Regulation of the immune response resulting from early exposure to those organisms no longer occurs, leaving the door open to an overly responsive immune system later in life.

In summary, there is evidence that our immune-related genes contain an especially high degree of variation, likely in response to strong historical selective pressures favoring resistance to numerous infectious diseases. Significant relaxation of natural selection has resulted from medical and public health advances that minimize our exposure to infectious diseases and their impact should we become ill. This may promote even greater genetic variation and allow for an increase in the frequencies of harmful variants embedded in our diversity of immune-related genes. As a result, genotypes that are no longer optimized to defend against pathogens may persist in our populations.

By eliminating the benefit of many immune genes, we are increasingly subject only to any adverse effects resulting from past evolutionary tradeoffs. Less frequent exposure to pathogens may leave some autoimmune disorders as the primary manifestation of genes or gene combinations that persisted previously due to their advantage in resisting infection. Because of this, we are experiencing an increasing burden of autoinflammatory and autoimmune disorders with significant underlying genetic risk factors. One contributing factor may involve our reduced early exposure to pathogens and to organisms with which we have coevolved, both playing a role in developing and regulating our immune response.

Addendum:

At the time of writing this, we are emerging from the worst pandemic in 100 years: the coronavirus disease known as Covid-19. Initially, without a vaccine or widespread immunity from previous exposure, human populations were profoundly susceptible to this infection. As of February 2023, the virus had been responsible for 674 million infections and over 6.86 million deaths worldwide[68], along with much morbidity and strain on healthcare systems. It resulted in profound changes in how we work, play, and socialize, with significant economic consequences. Although this appears to be a new human disease, it exemplifies what might be expected should some of our cultural adaptations, such as vaccines and antibiotics, become ineffective or unavailable. One of the most remarkable features of this pandemic is the extreme variability among individuals in the severity of response to this infection. Many people are essentially asymptomatic, while others suffer a protracted course of illness requiring intensive medical management, at times culminating in long-term disability or death. It was recognized early on that risk factors such as pre-existing disease or advanced age increase the likelihood of death.

There appears to be significant genetic variation in susceptibility to this disease. A viral epidemic in East Asia over 20,000 years ago has left a genetic signature of historical selection pressure on 42 genes coding for proteins that interact with the virus currently responsible for Covid-19[69]. The selected variants of these genes are found at intermediate frequencies, possibly accounting for some variability in response to the current pandemic. An analysis integrating the results of 46 GWAS involving nearly 50,000 patients from 19 countries identified thirteen genetic sites that were significantly associated with Covid-19 infection and/or severe manifestations[70]. Some of these sites are also known to be associated with lung or autoimmune inflammatory diseases. Another extensive genetic analysis of critically ill patients with Covid-19 documented 23 genetic variations that predispose them to critical illness[71]. They appear to be associated with the failure to control viral replication or an enhanced tendency toward inflammation in the lungs and blood clotting. Interestingly, one gene segment that is a major risk factor for Covid-19 appears to have been inherited from Neanderthals[72].

Our most effective tactics against this pandemic have been public health measures involving vaccines, wearing masks, social distancing, hand washing, assays for detection of the virus or antibodies to it, and contact tracing. It is unclear whether there will be significant consequences of natural selection related to this pandemic since most deaths have occurred in individuals of post-reproductive age. Regardless, the death and disruption we have experienced provide a real-time demonstration of how the mechanisms of natural selection may affect our lives.

Sources:

Aversa, Zaira et al. 2020. Association of infant antibiotic exposure with childhood health outcomes. Mayo Clinic Proceedings 96(1): 66-77.

Barber, Matthew F. et al. 2017. Rapid evolution of primate type 2 immune response factors linked to asthma susceptibility. Genome Biology and Evolution 9(6): 1757–1765.

Blaser, Martin J. 2017. The theory of disappearing microbiota and the epidemics of chronic diseases. Nature Reviews Immunology 17: 461-463.

Brinkworth, Jessica F. and Luis B. Barreiro 2014. The contribution of natural selection to present-day susceptibility to chronic inflammatory and autoimmune disease. Current Opinion in Immunology 31: 66–78.

Casanova, Jean-Laurent 2015. Severe infectious diseases of childhood as monogenic inborn errors of immunity. Proceedings of the National Academy of Sciences. www.pnas.org/cgi/doi/10.1073/pnas.1521651112.

Casanova, Jean-Laurent and Laurent Abel 2013. The genetic theory of infectious diseases: A brief history and selected illustrations. Annual Review of Genomics and Human Genetics 14: 215–243.

GBD 2019 Antimicrobial Resistance Collaborators 2022. Global mortality associated with 33 bacterial pathogens in 2019: a systematic analysis for the Global Burden of Disease Study 2019. The Lancet 400(10369): 2221-2248.

Karlsson, Elinor K. et al. 2014. Natural selection and

infectious disease in human populations. Nature Reviews Genetics 15: 379-393.

Lenz, Tobias L. et al. 2016. Excess of deleterious mutations around HLA genes reveals evolutionary cost of balancing selection. Molecular Biology Evolution 33(10): 2555–2564.

Okjabe, Hisao et al. 2023. Associations between fetal or infancy pet exposure and food allergies: The Japan Environment and Children's Study. PLoS ONE 18(3): e0282725. https://doi.org/10.1371/journal.pone.0282725.

Quintana-Murci, Lluis 2019. Human immunology through the lens of evolutionary genetics. Cell 177: 184-199.

Sanz, Joaquin et al. 2018. Genetic and evolutionary determinants of human population variation in immune responses. Current Opinion in Genetics & Development 53: 28–35.

Scudellari, Megan 2017. Cleaning up the hygiene hypothesis. Proceedings of the National Academy of Sciences 114(7): 1433–1436.

Siddle, Katherine J. and Lluis Quintana-Murci 2014. The Red Queen's long race: human adaptation to pathogen pressure. Current Opinion in Genetics & Development 29: 31–38.

Smallwood, Taylor B. et al. 2017. Helminth immunomodulation in Autoimmune Disease. Frontiers in Immunology 8: 452. https://doi.org/10.3389/fimmu.2017.00453.

Wibowo, Marsha C. et al. 2021. Reconstruction of ancient microbial genomes from the human gut. Nature 594: 234–239.

CHAPTER 12: Implications For Intelligence

Any attempt to explore social, cultural, genetic, or evolutionary issues regarding human intelligence is fraught with challenges[73]. Genetic and environmental effects are complex and difficult to sort out. The entire field of study suffers from the stigma of past attempts to demonstrate differences in intelligence based on "race." The measurement of intelligence itself is a point of controversy. Nevertheless, it is appropriate to address the topic here because of the degree of both scientific and general interest and the fact that it is likely to be impacted by human cultural evolution.

Intelligence is a complex quantitative trait. Although it is a concept that is generally familiar to all of us, it can be challenging to define precisely from a biological standpoint. Research studies often attempt to quantify intelligence using the measured Intelligence Quotient (IQ) or the general cognitive ability factor (g). IQ is a score derived from a specific set of standardized tests developed to measure a person's cognitive abilities. The g factor is more complex, derived from the correlations among performance on various cognitive tasks, that is, the degree to which performance on one task predicts performance on other tasks. This factor is multidimensional and incorporates the abilities to reason, plan, solve problems, think abstractly, comprehend complex ideas, learn quickly, and learn from experience. Another feature of g is that it reflects two types of intelligence; fluid and crystallized. Fluid intelligence refers to the capacity to think logically and solve problems in novel situations. We use fluid intelligence to solve puzzles, think abstractly, or invent new problem-solving strategies. Crystallized intelligence reflects our ability to use acquired information. It involves knowledge that comes from prior learning and past experiences.

Intelligence isn't just important for academic purposes; it has a real-life impact. It is associated with various social outcomes, including educational achievement, occupational success, income, and social mobility. In addition, people with higher intelligence tend to have better mental and physical health. They tend to choose healthier lifestyles with more exercise, a better diet, and less smoking, heavy drinking, or obesity. As a result, there is a lower risk of disorders such as cardiovascular disease, cerebrovascular disease, hypertension, and some cancers, as well as the likelihood of longer life expectancy.

Our cognitive abilities are one of the most defining features distinguishing humans from other species. It has been postulated that greater cognitive complexity associated with larger brains provided an evolutionary advantage by allowing us to cope more effectively with complex environments and to deal with evolutionarily novel situations and challenges. The latter would include those human-engineered changes that developed through cultural evolution and human niche construction. As we have seen, the ability of humans to create such changes lies at the heart of our success in circumventing the historical processes of natural selection. This ability, and the cognitive complexity it fosters, has been possible owing to the development of social skills that enable us to analyze the motivations of others and act in a cooperative, collaborative fashion. In humans, genes related to the development of nerve cell interconnections and cognitive function have been found to be enriched for evidence of recent positive selection. But just as we saw for the immune system, gene variants associated with central nervous system dysfunction, such as autism spectrum disorder or schizophrenia, also appear to have been subject to past positive selection. Despite their risks, those variants may have been favored due to their positive effects on nerve cell development, cognitive ability, and creativity. For example, the set of risk variants for autism spectrum disorder also correlates positively with years of schooling, college completion, childhood intelligence, and openness to experience.

When cognitive traits such as g are measured in a population, the result is a bell curve – with most individuals clustering around the average value and progressively fewer individuals as you approach the extreme high and low values. This variation among individuals is the result of both genetic and environmental effects. About 1-3% of the population will fall in the very lowest portion of the curve, where intellectual disability is characterized by poor performance on cognitive tests and difficulty functioning in daily life. The most common type of inherited intellectual disability is trisomy 21, also known as Down syndrome[74]. There are other recognized syndromes, and many individuals with cognitive impairments do not clearly fall into a recognized category, suggesting a polygenic basis. Over 300 specific genetic associations have been documented for intellectual disability, accounting for about 40% of cases.

The genetic component of the extensive variation in intelligence in the remainder of the general population also appears to be consistent with a polygenic, complex trait. Large genome-wide association studies on intelligence have suggested a role for hundreds of genetic variants, each with a rather small effect but cumulatively accounting for about 20-50% of the variation in general cognitive ability. Variants associated with cognition have been linked to processes such as brain cell development, myelination ("insulation" surrounding nerve fibers), communication among nerve cells, and regulation of the nervous system, as well as many more general biologic processes. A large multi-institutional study linked 1271 single nucleotide variants to cognitive traits and generated polygenic scores based on those variants to predict intelligence-related outcomes[75],[76]. They were able to explain 11-13% of the observed variance in educational attainment and 7-10% of variance in cognitive performance. A different multi-institutional study showed that polygenic scores based on cognition-related attributes like educational attainment could predict trends in childhood performance, such as earlier age of talking, enhanced self-control, and interpersonal skills[77]. Even beyond childhood, individuals with higher polygenic scores were found to be more geographically mobile, have more successful careers, secure mates of higher social status, and prepare better financially for retirement. Children from socio-economically disadvantaged families tended to have lower polygenic scores, but within that group, those with higher polygenic scores still tended to achieve more. The high degree of genetic variability in these processes provides considerable potential for evolutionary responses to the presence (or absence) of natural selection.

Although natural selection likely favored greater intelligence in the past, we now live in complex societies that often provide safety nets and other social structures that allow individuals from the lower half of the intelligence curve to succeed, or in evolutionary terms, to survive and reproduce. To some extent, if a society contains enough sufficiently intelligent individuals to serve the needs of government, science, medicine, and industry, all individuals in that society may reap the benefits, even those who may have been at a fitness disadvantage from a cognitive standpoint in past evolutionary times. There is no longer any reason to think that high cognitive function has as much of an adaptive advantage as it did previously. Stemming from these considerations, it has been predicted that a lack of ongoing positive or purifying selection for cognitive function, especially in combination with the increased rate of germline mutation associated with increased average parental age, will lead to a decline in average human intelligence. We know that neural structures or organs that no longer serve an adaptive purpose may undergo degeneration as new and previously harmful mutations accumulate due to the absence of selection against them. In addition, maintaining high levels of cognition may come at a cost to the organism since it consumes considerable metabolic energy and may, therefore, be selected against if that energy can be redirected to other functions that better serve to increase survival and reproductive success. It has been estimated that new mutations under a regimen of relaxed selection could account for a loss of average human IQ of nearly three points per generation.

The situation is complicated by additional factors that may influence the future course of human intelligence. Augmenting the effect of relaxed selection is a phenomenon termed *dysgenic fertility*. Dysgenic fertility refers to reproductive behaviors associated with lower intelligence that result in increased evolutionary fitness. In many developed and developing societies, individuals with lower intelligence reproduce at higher rates, on average, than those with higher intelligence. Individuals with higher intelligence tend to choose reproductive behaviors that put them at a disadvantage from an evolutionary standpoint. There are multiple reasons for this. Individuals with greater intelligence may be more likely to abandon conservative or religious values that favor fertility. The availability of more effective birth control methods provides women with the opportunity for extended education and increased participation in the workforce. There may even be an ethical desire to limit global population growth. In general, the effect is more pronounced for women than men. As stated by Dr. Hannah Peach, a psychologist at the University of North Carolina Charlotte, and her colleagues:

> *the control of reproduction can be thought of as an evolutionarily novel event...since higher levels of g are posited to reflect a greater ability to navigate evolutionarily novel tasks, this would suggest that individuals higher in g can more successfully control reproduction.[78]*

The negative relationship between fertility and intelligence, socioeconomic status, or education (the latter two traits being independently correlated with intelligence) extends back to the early 19th century in many developed countries and has been documented more recently in the United States, Great Britain, Australia, China, Iceland, India, Turkey, Russia, Denmark, Sweden, and Finland, as well as in multinational datasets. The previously mentioned study at the University of North Carolina Charlotte has examined a database of over 325,000 high school students in the US who were followed for up to 11 years post-graduation. They found that both greater intelligence (g) and socio-economic wealth independently predicted higher educational attainment, delaying marriage and reproduction, and having fewer total offspring. Their results would imply a loss of about 0.5 to 1.5 IQ points per generation, depending upon sex and the duration of follow-up. Independently, Gerhard Meisenberg from Ross University Medical School in Dominica has estimated that dysgenic fertility may be associated with a decrement of 1.63 points of IQ per generation and that absent this effect in recent history, average human intelligence would currently be as much as five points higher[79]. He goes on to predict a future decline of about 2.9 points per century, and the proportion of gifted people with IQ above 130 could decrease by 37.7%. Michael Woodley of Menie from the Technische Universität Chemnitz in Germany has estimated that the adverse fitness effects of dysgenic selection combined with those of parental age and accumulated mutation load could account for a loss of .53 to 1.53 points of heritable g per decade, or 1.86 to

6.72 points per generation[80].

Detection of potential early evolutionary effects of dysgenic fertility is complicated by an opposite trend in intelligence that has been termed the Flynn effect (named for James R. Flynn, who was one of the first to document this trend): a consistent finding of increased measured intelligence of about 3 points per decade in populations throughout the 20th century. It has been attributed to non-genetic environmental factors such as better health, more and better education in school and at home, better-educated parents, and rising standards of living. The Flynn effect will likely mask any negative impact of dysgenic fertility. However, there may be a limit to the degree of continuing benefit that can accrue from improvements in our environment. In fact, there is accumulating evidence for trends toward a weakened Flynn effect or even a reversal. A decline in IQ has been documented for populations from France, Norway, Australia, Denmark, United Kingdom, Sweden, Netherlands, and Finland, with the decrease ranging from 0.26 to 4.3 IQ points per decade. The relative negative impact appears to be greatest among those with the highest cognitive performance. In addition, there is evidence that other parameters linked to intelligence - such as processing speed, working memory, and innovation - have declined in Western countries. The causes of this "negative Flynn effect" are uncertain, but it has been attributed to environmental factors, changes in the composition of subject populations because of immigration, or dysgenic fertility. There is some evidence to indicate that the decline is genetically based. For example, a group at the University of Iceland has documented a slow but steady decline in the frequency of certain variants associated with educational attainment throughout the 20th century, presumably the result of negative selection resulting from dysgenic fertility[81].

The effects of a collective decline in intelligence could be profound for individuals and their societies. There may be a compounding effect: declining intelligence might render populations less able to maintain optimal environments (e.g., failure of niche construction), so our ability to avoid natural selection may decline along with our intellect. The pace of technological and scientific innovation could also be impacted. There are implications for our health owing to its positive association with intelligence as well. Social outcomes such as political involvement, liberal attitudes, and social progress could be impacted. Ultimately, national prosperity and government effectiveness could decline in the face of a world increasingly challenged by the complexity of modern technology, global warming, pandemics, social unrest, and demographic pressures.

These outcomes are not inevitable. Dysgenic fertility is driven by individual behaviors that are malleable and influenced by cultural and environmental factors. Therefore, it is possible that the magnitude and direction of the intelligence/fertility relationship may change over time so that intelligence is again favored, or at least not selected against. In Nordic countries, for example, childlessness has increased among the lesser educated, and the negative relationship between women's education and fertility is no longer as clear.

> *"If education is taken as a proxy for a person's earnings potential, having a sufficiently high income or having the economic means to*

> *sustain a family may have become an increasingly important prerequisite for having (more) children for both women and men...the difficulties reconciling a career with family building that previously hampered the childbearing of highly educated women, in particular, seem to have been overcome[82]."*

In the United States, there is evidence that fertility has increased among highly educated women. A recent study found the relationship between fertility and women's education in the US to be U-shaped, with fertility increasing at both lower and higher educational levels, the latter presumably due to offloading of time-consuming demands to services such as childcare and housekeeping[83]. However, peak reproduction of the most highly educated women was still delayed by about ten years, which suggests ongoing negative consequences for fitness.

In summary, human intelligence has likely evolved as an adaptation to optimize fitness in complex, novel environments – including those highly crafted by humans themselves. Intelligence has profound implications for both individuals and the societies within which they live. Apart from specific genetic abnormalities associated with recognized syndromes, most human cognitive variation results from a complex interplay of environmental factors and considerable polygenic variation, providing the potential for ongoing evolution. In the setting of relaxed selection, mutations with adverse effects on intelligence may accumulate. Interpreting current trends in human intelligence is difficult due to dynamic social/cultural influences such as dysgenic fertility and the Flynn effect. There is a risk that the long-term effects of relaxed selection on intelligence may not become apparent until they are well underway.

Sources:

Bender, Andrea 2019. The role of culture and evolution for human cognition. Topics in Cognitive Science 2019: 1–18.

Davies, Gail et al. 2018. Study of 300,486 individuals identifies 148 independent genetic loci influencing general cognitive function. Nature Communications 9:

2098. https://doi.org/10.1038/s41467-018-04362-x.

Deary, Ian J. et al. 2019. What genome-wide association studies reveal about the association between intelligence and physical health, illness, and mortality. Current Opinion in Psychology 27: 6–12.

Flynn, James R. and Michael Shayer 2018. IQ decline and Piaget: Does the rot start at the top? Intelligence 66: 112–121.

Kanaya, Tomoe 2016. Discussing the Flynn effect: From causes and interpretation to implications. Measurement: Interdisciplinary Research and Perspectives 14(2): 67-69.

Lynn, Richard et al. 2018. Regional differences in intelligence in 22 countries and their economic, social and demographic correlates: A review. Intelligence 69: 24–36.

Malykh, S. B. et al. 2019. Molecular genetic studies of cognitive ability. Russian Journal of Genetics 55(7): 783–793.

Plomin, Robert and Sophie von Stumm 2018. The new genetics of intelligence. Nature Reviews Genetics 19: 148-159.

Sauce, Bruno and Louis D. Matzel 2018. The paradox of intelligence: Heritability and malleability coexist in hidden gene-environment interplay. Psychological Bulletin 144(1): 26-47.

Woodley of Menie, Michael A. et al. 2015. Estimating the strength of genetic selection against heritable g in a sample of 3520 Americans, sourced from MIDUS II. Personality and Individual Differences 86: 266–270.

SECTION FIVE: OUR GENETIC FUTURE

CHAPTER 13: The Path Forward

"Having carved out our own strange niche, far removed from the laws that govern the rest of the natural world, we have only ourselves to look to for guidance."[84]

In this book, I have laid out the rationale and evidence that raise concerns about our genetic future. As I stated at the outset, I view the problems of human evolution both as an evolutionary biologist and a physician. From the medical standpoint, let me be very clear; we run the risk of circumventing the differential survival and reproduction that underlies natural selection and the maintenance of our genome whenever we:

- perform cardiopulmonary resuscitation
- place someone on a mechanical ventilator
- treat infection with antibiotics
- introduce a new vaccine
- operate to remove a tumor
- medically assist reproduction
- perform a Caesarian section for a placental abruption

- utilize any of the numerous other medical interventions

There is a strong argument here for finding ways to prevent or avoid the consequences of gradual genetic decline. Reinstating some degree of natural selection by withholding treatment of those facing death or reproductive failure is not ethically acceptable. There is no simple remedy, but three possible strategic approaches merit further consideration.

Improvements in Disease Management

We have made significant progress in understanding the nature and extent of genetic variation among individuals. Though, we still have a long way to go before we understand the function of much of this variation and can translate that knowledge into truly personalized medicine: employing individually tailored prevention and treatment strategies that consider each person's unique genetic makeup. Coupled with emergent medical technologies such as those discussed in this chapter, personalized medicine can go a long way toward improving the treatment of afflicted individuals.

Among the many new developments in medical therapeutics, none has more potential to revolutionize our approach to patient care than our evolving ability to modify genes to prevent or treat disease. Gene therapies offer the possibility to correct or cure the underlying problem rather than treat the manifestations of the disease. This principle is illustrated by current therapies for a devastating disorder called spinal muscular atrophy that affects around 400 babies annually in the US. It is caused by a defective gene that results in severe muscle weakness. These infants are so weak that they are ultimately unable to move, swallow, or breathe, resulting in death before the age of two. The gene normally codes for a protein designated SMN that is essential for maintaining motor neurons, cells that relay signals from the brain and spinal cord to "tell" the skeletal muscles to contract when we move. When the gene is defective, the lack of SMN protein results in the death of motor neurons, causing muscle wasting and weakness. There is some good news. A remarkable new therapy approved in 2019 employs a functional replacement gene synthesized in the laboratory and delivered to the cells by a virus vector. This technique utilizes a benign virus whose intrinsic DNA has been replaced with the synthesized gene. Once administered to a patient, the virus attaches and injects the new DNA into the motor neuron cells, restoring their ability to synthesize the SMN protein. Because it restores the cells' normal function, this approach is considered curative with one course of treatment. It is not all good news, however, as the patient's offspring do not inherit the corrected gene, and the risk of transmitting the genetic defect to subsequent generations still exists. In addition, the cost of treatment is $2.125 million. Costs of this magnitude on any larger scale would quickly become unsustainable, especially in the face of the increasing overall genetic burden to society over time.

A different genetic technique is employed with a drug called nusinersen, which has also been approved to treat spinal muscular atrophy. This drug capitalizes on a second gene humans carry that also produces the SMN protein, but only about 10% of the protein it produces is normally functional. The problem involves an error in the RNA editing process. When DNA that codes for a protein is "read" or transcribed, some chemical editing of the resulting RNA must occur before the final instructions are ready for use in protein synthesis. Some segments of the RNA molecule are not meant to be translated into code for the protein, and they must be removed. This editing process is faulty in the case of the second SMN gene. Nusinersen binds to RNA and corrects the editing process so that the gene can produce a fully functional SMN protein. The drug must periodically be injected directly into the central nervous system to be effective. The current treatment cost for this drug is $750,000 in the first year and $375,000 annually after that.

Viral delivery of gene therapy is also utilized by a new drug marketed as Hemigenix, approved by the U.S. Food and Drug Administration in November 2022 for treating hemophilia B. Hemophilia is a genetic bleeding disorder that occurs when certain blood factors required for clotting are absent. Hemophilia A (missing factor VIII) is more common, while hemophilia B (missing factor IX) accounts for about 15% of cases. People with hemophilia are subject to uncontrolled bleeding, which can sometimes be fatal. The new treatment involves the modification of a virus to deliver a fabricated gene to the patient's liver cells. This gene codes for the factor IX protein and restores some degree of normal clotting ability, eliminating the need to take frequent and expensive prophylactic doses of factor IX in nearly all patients[85]. This one-time treatment costs $3.5 million, making it the most expensive drug in the world.

Sickle cell disease is one of the most common disorders resulting from a recessive mutation in a single gene, affecting around 100,000 people in the US. We previously saw how this gene is involved in the synthesis of hemoglobin, an essential blood-cell protein that carries and delivers oxygen to all the body tissues. When the gene carries two copies of the mutation, the structure of all the hemoglobin in red blood cells is altered. The abnormal hemoglobin causes the cells to become deformed into a "sickle" shape. They may clump together and clog blood vessels, depriving tissues of oxygen and causing episodes of severe pain. Several clinical trials are underway to explore different sorts of genetic interventions to reduce the impact of this disease. One trial takes advantage of the fact that humans normally have two types of hemoglobin. Fetal hemoglobin is produced in the fetus, but its production shuts down by about six months of age due to the action of another gene designated BCL11A. At that point, the adult form of hemoglobin takes over, which is the type affected in sickle cell disease. The new therapy involves removing the patient's stem cells that are involved in the production of red blood cells. Part of the BCL11A gene in these cells is removed using a popular gene-editing process called CRISPR–Cas9, then the cells are reintroduced to the patient. In the absence of a functioning BCL11A gene, the production of fetal hemoglobin is not blocked. Instead, the persistent fetal hemoglobin compensates for the defective adult form and relieves the debilitating painful crises.

A unique approach has been developed using genetic engineering to fight certain cancers. T cells are a type of white blood cell with immune functions. Some of them work by binding to infected cells and killing them. The new anti-cancer therapy involves "reprogramming" T cells to recognize and kill cancer cells. First, the patient's own T cells are removed from their blood. Then, they are modified using a harmless virus to introduce new genetic instructions allowing them to recognize, target, and kill cancer cells[86]. Finally, the modified T cells are multiplied in culture and injected back into the patient; they are sometimes called a "living drug," which is specific for each patient and durable since the cells survive and proliferate inside the patient.

Another remarkable approach has been approved in the UK for preventing disorders such as Leber hereditary optic neuropathy, a form of blindness caused by mutations in mitochondrial DNA. Apart from the nucleus, mitochondria are the only other structures within cells that contain heritable DNA. They are only inherited through the mother since they occur in the egg cells, not the sperm. So, an egg cell can have normal DNA in the nucleus but disease-related mutations in the mitochondrial DNA. In mitochondrial replacement therapy, the normal maternal nucleus from an egg with faulty mitochondria and all the normal DNA within it is transferred to replace the nucleus from a healthy donor egg with normal mitochondria. After fertilization, the individual that results will actually contain DNA from three "parents": the nuclear DNA from both the mother and the father and mitochondrial DNA from a donor female.

A recent development has been the therapeutic use of manufactured RNA molecules. A dramatic example has been the rapid development and administration of RNA that instructs the subject's cells to produce protein sequences found on the exterior of the coronavirus that causes COVID-19. The protein itself is harmless, but it triggers the immune system to produce antibodies that work against the virus and are very effective at preventing severe disease.

Evolving gene therapies can potentially treat the underlying causes of many simple, genetically based disorders. Reproductive technologies such as gametal or embryonic genetic engineering may be one of the more powerful approaches to dealing with harmful genetic variants. They also provide the added benefit of allowing for the transmission of "corrected" genomes to an individual's offspring. In fact, a team in China has genetically engineered human embryos with a mutation that is thought to reduce the chances of HIV infection. The engineered embryos were then implanted successfully, resulting in the birth of twin girls[87]. Other technologies include the use of pluripotent stem cells, special human cells that can potentially develop into all the different types of adult cells. Current studies explore their possible use to treat Parkinson's disease, blindness, and spinal cord injuries. They could also be used to create human eggs and sperm in a culture dish which would facilitate the screening and selection of gametes or embryos prior to implantation to avoid unfavorable genetic characteristics.

None of these emerging interventions is without potential problems. First, most of them leave a person's original germ cells intact so that the genetic basis for disease continues to be inherited. The CRISPR–Cas9 tool used for gene editing may produce errors, although newer methods with greater precision and reliability are being developed. Unintended effects could result when genes affect more than one trait; seemingly harmful variants that are modified may have other hidden beneficial effects that could be adversely affected. The effects of an individual gene may also vary depending upon the specific array of other gene variants also carried by that person, as well as any number of environmental variables. Even though gene therapies have significant potential to modify or even cure monogenic disorders, they may have very limited use for most common human diseases, which are complex. For those diseases, the degree of risk may involve hundreds or thousands of genetic variants, each of minor effect. Finally, the cost of these therapies can be exorbitant and may limit their availability to some patients who could otherwise benefit, or place an untenable financial burden on the entire healthcare system.

Public Health Approaches
The issues of relaxed selection and increased genetic load have largely been absent from public health policy considerations. However, public health genomics has been established as a relatively new field of study aimed at using genetic information to improve the health of populations. Within this context, genetic issues can be assessed regarding their impact on contemporary public health and implications from an evolutionary perspective. The goals of this approach include assessment of individuals' genetic risk for disease, increased attention to the gene/environment interactions that contribute to disease, and targeted public health interventions.

Advances in genetic screening, even before conception, provide new possibilities for identifying and proactively modifying the negative consequences of harmful genetic variants. One study looked at a screening for simple recessive mutations known to cause genetic disease in over 23,000 individuals. Twenty-four percent of individuals were carriers for at least one of 108 disorders, and 5.2% were carriers for multiple disorders.

As part of in vitro fertilization, embryos can now be screened for Tay-Sachs disease, cystic fibrosis, and phenylketonuria so that affected embryos can be rejected for implantation.

Screening for complex diseases is more challenging. One active area of research involves the determination of polygenic risk scores (PRS). As we have seen, complex diseases often reflect the overall impact of numerous genetic variants, each with a relatively small effect. Previously, we have only been able to approximate genetic risk by assessing the individual's family history. Now, we can sequence DNA from individuals, and ongoing genome-wide association studies contribute to a growing database of risk estimates for individual genetic variants. The PRS incorporates information about the particular spectrum of variants carried by an individual and estimates the overall increased risk of disease based on their combined effect. Thus, PRS's can provide direct genetic risk information in addition to and independent of family history. For example, risk scores have been shown to identify individuals most likely to benefit from statin therapies used to reduce serum cholesterol. One such study found that patients in the top 20% of risk based on PRS's have a greater likelihood of a coronary artery disorder, and they can reduce the risk of heart attack or death by about 45% upon initiating statin therapy. PRS's could also help inform decisions about how and when to screen for some diseases and provide information for patients about the likelihood of future diseases that may suggest preventive interventions or assist with life-planning decisions. Another study examined five common diseases in populations of European ancestry: coronary artery disease, atrial fibrillation (an abnormal heart rhythm), type 2 diabetes, inflammatory bowel disease, and breast cancer. The proportion of the population found to be at threefold increased risk for any of these diseases based on PRS's ranged from 1.5 to 8%. Awareness of the elevated risk of disease in those individuals can facilitate early intervention to prevent or moderate the impact.

Polygenic risk scores also have significant limitations. For one, they are only as good as the GWAS data used to provide risk estimates for individual genetic variants. The greatest concern is that the vast majority of GWAS's have been performed on human populations from Europe or their descendants (e.g., many US populations). Applying these data to populations with different genetic ancestry, African, for example, may be very misleading since their profile of genetic variants and their relative risks, as well as their environments, can be quite different. Genetic susceptibility is only one component of disease risk. PRS's do not consider the two other primary components: environmental exposures and lifestyle factors.

Despite their limitations, PRS's are coming into use and are even being marketed directly to consumers[88]. Unfortunately, there is little to no oversight, raising the possibility that consumers will access this information while having limited insight into the potential problems and risks of interpreting such complex genetic data. Some companies even offer prospective parents complex screening tests on embryos as part of the in vitro fertilization procedure. The presumed risk of future complex diseases may then be used for embryo selection, despite the fact that there are no scientific studies validating this practice. There are also numerous pitfalls in attempting to apply genetic risk estimates to embryos when those estimates are based on GWAS's of adults with limited ancestry (e.g., European) or living in different environments[89].

At the forefront of current developments in personalized medicine is an attempt to improve individual screening for current and future disease risks. A concept generally known as phenomics begins with the assessment of genetic risk but incorporates many other biological, environmental, and lifestyle measures to provide a comprehensive analysis of one's state of health at a given time as well as the magnitude of future health risks. One version of this approach aims to include the following assessments:

1. Complete genome sequencing
2. Analysis of thousands of proteins
3. Evaluation of the gut microbiome
4. Analysis of thousands of metabolites - small molecules reflecting biochemical activity
   in the human body.
5. Brain health, based on imaging, surveys, and examination
6. Social and lifestyle, including environmental exposures and traumas
7. Vital signs, blood tests, and other standard evaluations

Some form of advanced artificial intelligence would be necessary to process all this information, relate the results to known medical risks, and produce a "phenomics report card" that will guide physicians and their patients regarding interventions to prevent or prepare for future health problems.

Strategic Genetic Choice

In 1950, the eminent geneticist Hermann Joseph Muller addressed the problem of increasing mutation load, noting that either the mutation rate must be decreased somehow or some form of selection reinstituted, suggesting that voluntary selection is a necessary complement to medicine. He hoped that with proper education and guidance, individuals laden with harmful mutations, assuming they can be identified, would choose to avoid reproduction. Similar concerns were addressed by the World Health Organization in 1972. They recommended the establishment of medical genetics centers, educational initiatives, prenatal diagnosis, registries of genetically determined disorders, and research on the impact of interventions on the frequency of genetic diseases. Susan Holloway and Charles Smith of the University of Edinburgh, Scotland, wrote about possible specific practices in 1975[90]. These included: (1) improved treatment of affected individuals, (2) voluntary selection of mates based on genetic screening for harmful recessive traits that might affect offspring if both partners carry the same mutation, (3) selective voluntary abortion of genetically impaired fetuses with or without further attempts at reproduction, (4) voluntarily limited reproduction by carriers of harmful recessive traits, particularly those posing a greater risk to offspring, and (5) artificial insemination by selected donors in couples where the male is at risk of passing on harmful variants. In general, these approaches aimed to achieve the benefits of natural selection for limiting the spread of disease-related gene variants, but without the suffering and death that restoration of natural selection would entail.

Many people are uncomfortable with the concept of choosing which genetic lineages should be discouraged from reproducing or which pregnancies (if any) should be terminated. The practice of abortion itself remains highly controversial. Furthermore, these discussions raise the specter of ethically misdirected approaches to eugenics from the 20th century – including involuntary sterilizations, marriage restrictions, and genocide. Despite the negative connotations of the term, eugenics can be defined as the study (and ultimately the practice) of arranging reproduction within human populations to increase the occurrence of heritable characteristics regarded as desirable. In the current context, minimizing human disease is undoubtedly desirable, but the concept is still problematic.

From an ethical standpoint, imposing differential survival and/or reproduction based on genetic or other characteristics remains unacceptable. But some authors have argued that advances in medical technology are already ushering in a "second age of eugenics" or "velvet eugenics," although on a voluntary basis. Advances in genetic and reproductive techniques already provide greater opportunities for preventive genetic medicine, with individuals having more information and choices regarding their own reproduction. Newer non-invasive pre-conception, embryonic, and fetal screening tests already available raise the possibility of reducing the prevalence of many simple genetic diseases. The information available from polygenic risk scores, should they become validated, may also provide a basis for couples to make decisions about reproduction.

Genetic testing and voluntary abortion have already significantly impacted the prevalence of some genetic syndromes. Denmark was one of the first countries to offer prenatal screening for Down Syndrome, which results from the inheritance of an extra chromosome #21. Nearly all pregnant women in Denmark choose to undergo this test, and of those found to have a fetus with Down Syndrome, more than 95% chose to have an abortion. As a result, in 2019, only 18 babies with Down syndrome were born in Denmark. Another example is Tay-Sachs disease, which causes severe neurological impairment and death resulting from a recessive genetic mutation, especially prevalent in Ashkenazi Jewish populations. A combined approach of genetic screening to discourage marriage between carriers of the recessive gene, pre-implantation testing of embryos, and prenatal fetal testing have nearly eliminated this disease.

In some cases, the ability to choose the genetic characteristics of our offspring has risen to the status of an individual right, at least in some people's minds. For example, a Southern California couple planning a pregnancy wanted to avoid passing along disease-related variants they carried to their offspring. The woman carried the BRCA1 mutation associated with an increased risk of breast cancer, while the man carried a rare mutation that confers around an 80% lifetime risk of experiencing a form of stomach cancer called hereditary diffuse gastric cancer (HDGC). This cancer tends to occur at a young age and is difficult to detect early, so it is often advanced by the time it is diagnosed; the best prevention is to undergo complete surgical removal of the stomach. Because of their genetic concerns, this couple opted to pursue IVF with genetic testing of the embryo before implantation. But after a successful birth, it was discovered that the embryo had inherited the

HDGC mutation and that this information appeared to have been covered up by the fertility clinic. The couple is now arbitrating a claim due to the wrong embryo being implanted and are suing the clinic, their doctor, and IVF coordinator for fraudulent concealment. The mother stated: "We trusted them to help us have a healthy baby"[91].

Ethical concerns have also been raised about our ability to limit genetic techniques to minimize disease risk when similar approaches may provide the potential for making reproductive choices solely based on personal preferences or a misguided desire to improve the human genome. Many people are choosing to undergo genetic profiling as an aid to discovering their genealogy, and it is easy to imagine a time in the future when the decision to marry may be influenced by the analysis of a complete genetic profile of the prospective partners. Other couples with difficulty conceiving may turn to assisted reproduction, a booming and largely unregulated business. Prospective parents seeking donor eggs or sperm can search online and choose from any number of characteristics, including donor build, intelligence, education, race, ethnic background, talents, and even face-matching with prospective parents or with celebrities. The American historian Nathaniel C. Comfort stated that "molecular genetics has significantly changed what counts as a disease. Doctors are increasingly moving from treating disease to treating risk...In the end, no logical friction can slow the slide from prevention to enhancement"[92]. Such "designer babies" will have the chosen heritable characteristics in their DNA. Are we really ready to engineer our own evolution? As posed by Eric S. Lander, the director of the Broad Institute of MIT and Harvard, "Would we come to regard our children as manufactured products? Would marketers shape genetic fashions? Would the "best" genomes go to the most privileged? If we cross this threshold, it's hard to see how we could ever return[93]."

Conclusion

For better or for worse, intentionally or not, we continue to make choices that have the potential to alter the genetic composition of future generations, and those choices will have implications both for the individual and for society. We need to recognize that the effects may not always be concordant, leading to conflicting priorities. The conversations that result may be uncomfortable, but that is no reason to ignore the reality of the genetic challenges we face. Advancing medical/genetic technology will require a broader perspective, with ongoing education and careful communication between physicians and their patients. Equally important will be determining what is considered ethically acceptable and what is not. At present, an approach combining advances in medical treatment, public health interventions, and rational reproductive choices based on risk assessment may serve to ameliorate, to some extent, the adverse impact the relaxation of natural selection is having upon our genetic well-being. We will need to continue supporting the lives and well-being of individuals who carry sub-optimal genotypes associated with disease or traits falling in the lower tail of the bell curve of "normal" human function. Economic and social support systems should be strengthened and expanded. It is imperative to the future of humanity that we recognize the potential hazards we face and work to find novel and ethically impeccable ways to address our ongoing genetic challenges.

Sources:

Duncan, David Ewing 2022. The Phenomics Revolution. Scientific American Health December 7, 2022. https://www.scientificamerican.com/custom-media/the-new-science-of-wellness/the-phenomics-revolution/

Eisenstein, Michael 2021. Fix the gene, cure the disease. Nature 596 Supplement: S2-S4.

Forzano, Francesca et al. 2022. The use of polygenic risk scores in pre-implantation genetic testing: an unproven, unethical practice. European Journal of Human Genetics 30: 493–495.

Gallagher, James 2019. 'Living drug' offers hope to terminal blood cancer patients. BBC News: Health, June 21, 2019. https://www.bbc.com/news/health-48706822.

Hodson, Richard 2021. Stem cells. Nature 597 Supplement: S5.

Khera, Amit V. et al. 2018. Genome-wide polygenic scores for common diseases identify individuals with risk equivalent to monogenic mutations. Nature Genetics 50: 1219–1224.

Kim, Michelle S. et al. 2018. Genetic disease risks can be misestimated across global populations. Genome Biology 19: 179. https://doi.org/10.1186/s13059-018-1561-7.

Kingwell, Katie 2023. First CRISPR therapy seeks landmark approval. Nature Reviews Drug Discovery News, April 3, 2023. https://doi.org/10.1038/d41573-023-00050-8.

Kumar, Akash et al. 2022. Whole-genome risk prediction

of common diseases in human preimplantation embryos. Nature Medicine 28: 513-516.

Lander, Eric S. 2015. Brave new genome. New England Journal of Medicine 373(1): 5-8.

Li-Pook-Than, Jennifer and Michael Snyde 2013. iPOP goes the world: Integrated personalized omics profiling and the road toward improved health care. Chemistry & Biology 20: 660-666.

Munsie, Megan and Christopher Gyngell 2018. Ethical issues in genetic modification and why application matters. Current Opinion in Genetics & Development 52: 7–12.

Naddaf, Miryam 2022. Scientists welcome $3.5-million drug – but questions remain. Nature 612: 388-389.

Rand, Leah Z. and Aaron S. Kesselheim 2020. Million-Dollar Drugs. Scientific American Custom Media; Health, Ethics and Innovation: 17-19.

Smith, Shelley L. 2016. Twenty-first century "eugenics"?: The enduring legacy. Perspectives in Biology and Medicine 59(2): 156-171.

Sufian, Sandy and Rosemarie Garland-Thomson 2021. The Dark Side of CRISPR. Scientific American, February 16, 2021. https://www.scientificamerican.com/article/the-dark-side-of-crispr/.

Torkamani, Ali et al. 2018. The personal and clinical utility of polygenic risk scores. Nature Reviews Genetics 19: 581–590.

Zhang, Sarah 2020. The Last Children of Down

Syndrome. The Atlantic 36(5): 42-55.

GLOSSARY

Adaptation.
The process of adjusting behavior, physiology, or structure to become more suited to an environment, particularly with regard to survival and reproduction. The word also refers to the outcome of that process.

Adaptive immune system.
Part of the immune response involving specialized immune cells that recognize and destroy specific foreign invaders. It prevents recurrent disease by retaining a memory of the antigens involved and mounting a rapid immune response. Adaptive immunity may last for a few weeks or months or sometimes for a person's entire life.

Agricultural transition.
Also called the Neolithic Revolution, it refers to a time about 10,000 to 12,000 years ago when traditional hunter-gatherer lifestyles, followed by humans since their evolution, were replaced with permanent settlements where crops and animals could be farmed to meet demand.

Alleles.
Alternate versions of a gene due to differences in the base sequence they contain. An individual inherits two alleles for each gene, one from each parent.

Amino acids.
Small molecules that are the building blocks of proteins.

There are twenty different amino acids making up human proteins, and each is represented by a triplet of three DNA bases in the genetic code.

Antagonistic pleiotropy.
When a gene variant affects more than one phenotypic trait, with opposite effects on fitness.

Antibodies.
Proteins in the blood produced by plasma cells (a type of white blood cell) in response to specific antigens associated with bacteria, fungi, viruses, or toxins. They bind to the antigen as part of a process leading to destruction and removal of the harmful substance. Part of the adaptive immune response.

Antigens.
Any substance that induces the immune system to produce antibodies against it.

Assisted reproductive technology (ART).
Any of several medical procedures used primarily to address infertility, especially treatments in which eggs or embryos are manipulated. Most commonly, eggs are surgically removed from a woman's ovaries, combined with sperm in the laboratory, and returned to the woman's body or donated to another woman.

Autoimmune disease.
Malfunction within the adaptive immune system when the body's natural defense system can't tell the difference between your own cells and foreign cells, causing the body to mistakenly attack normal cells.

Autoinflammatory disease.
Unprovoked activation of the innate immune system producing symptoms such as fever, rash, or joint swelling that mimic active infection.

Black Death.
A pandemic caused by the plague bacterium *Yersinia pestis* that occurred in Western Eurasia and North Africa from 1346 to 1353. It is the most fatal pandemic recorded in human history, causing the deaths of 75–200 million people. Humans usually get plague after being bitten by a rodent flea.

Bottleneck.
When a population experiences a marked reduction in size, which may be followed by renewed growth.

Cephalopelvic disproportion.
Mismatch between the size of the fetal head (too large) and size of the mother's birth canal (too small), resulting in "failure to progress" in labor and potentially endangering the lives of both mother and fetus.

Chromosome.
A chromosome is a thread-like structure found inside the nucleus of a cell that contains genetic information in the form of DNA organized into genes. Human cells normally contain 23 pairs of chromosomes.

Coevolution.
When two or more organisms reciprocally adapt through natural selection in response to the long-term presence of the other organisms in its environment. A common example involves the mutual adaptation of flowering plants and their pollinators.

Complex disease.
A disorder that results from the contributions of multiple genes and gene variants along with significant influences of the physical and social environment. The presence of a particular gene variant in an individual contributes to the risk of disease but is not in itself enough to cause

the disease. Common complex diseases include obesity, cardiovascular disorders, chronic obstructive pulmonary disease, and type 2 diabetes.

Culture.
All the ways of life including arts, beliefs, institutions, and other manifestations of human intellectual achievement that are socially transmitted and passed down from generation to generation.

Demographic transition.
A long-term trend of declining birth and death rates in a population, often associated with economic improvement.

DNA, Deoxyribonucleic acid.
The molecule that carries genetic information for the development and functioning of an organism. DNA is made of two linked strands that wind around each other to resemble a twisted ladder — a shape known as a double helix. Each strand has a backbone made of alternating sugar (deoxyribose) and phosphate groups. Attached to each sugar is one of four bases: adenine (A), cytosine (C), guanine (G) or thymine (T). The sequence of the bases along DNA's backbone encodes biological information, such as the instructions for making a protein or RNA molecule.

Dominance.
Refers to the situation where a gene has two variants, or alleles. If one allele masks the effect of the other (which is recessive) when both are present, it is said to be dominant.

Dysgenic fertility.
The negative correlation between intelligence and number of children, so that reproductive behaviors

associated with lower intelligence result in increased evolutionary fitness.

Eugenics.
Has been variously defined, but in a broad sense refers to the study (and ultimately the practice) of how to arrange reproduction within human populations to increase the occurrence of heritable characteristics regarded as desirable.

Evolution.
Changes in the heritable traits of a population over successive generations.

Fertility.
The production of healthy offspring in abundance, determined by both physiology and behavior.

Fitness.
The relative ability of organisms to survive and reproduce in their environment and contribute genes to the next generation.

Flynn effect.
The increase in measured intelligence in populations throughout the 20th century. It has been attributed to non-genetic, environmental factors such as better health, more and better education in school and at home, better educated parents, and rising standards of living.

Founder effect.
The reduction in genetic variation that results when a new population is established by a very small number of individuals from a larger population. The **serial founder** model suggests that human expansion generally involved numerous bottlenecks as small bands of individuals repeatedly pushed into unoccupied territories at the expanding front of human occupation.

Frequency-dependent selection.
A situation where fitness is dependent upon the frequency of a gene variant in a population. In some cases, the fitness of a variant increases as its frequency in a population decreases.

Gametes.
The reproductive cells that contain half of the genetic content of the mature individual. Female gametes are called egg cells, and male gametes are called sperm.

Gene.
A specific segment of DNA (deoxyribonucleic acid) that contains information needed to influence specific physical or biological traits. Many genes code for specific proteins whereas others serve to modify the activity protein-coding genes.

Gene expression.
The result of a gene becoming activated to make RNA molecules that may code for proteins or serve other functions.

General cognitive ability factor (g).
A metric of intelligence that is derived from the correlations among performance on a variety of cognitive tasks, that is, the degree to which performance on one task predicts performance on other tasks. This factor is multidimensional and incorporates the abilities to reason, plan, solve problems, think abstractly, comprehend complex ideas, learn quickly, and learn from experience.

Genetic architecture.
The total of all genetic contributions to a particular trait. It is influenced by the number of genetic variants affecting a trait, their frequencies in the population, the

magnitude of their effects and their interactions with each other and the environment.

Genetic code.
The instructions contained in a gene that tell a cell how to make a specific protein. Each gene's code uses the four nucleotide bases of DNA: adenine (A), cytosine (C), guanine (G) and thymine (T) — in various ways to spell out three-letter "codons" that specify which amino acid is needed at each position within a protein.

Genetic variation.
The difference in DNA base sequences between individuals within a population. Alternate variants at a single gene are called alleles.

Genome.
The complete set of DNA (genetic material) in an organism.

Genome-wide association study (GWAS).
Studies that test hundreds of thousands of genetic variants across many genomes to find those that are statistically associated with a specific trait or disease.

Genotype.
The complete, heritable, genetic makeup of an individual that contributes to their phenotype.

Germline.
The cells that form eggs in females and sperm in males. Germline cells contain the genetic information that is passed down from one generation to the next.

Hemoglobin.
A protein in red blood cells that carries oxygen to the body's organs and tissues and transports carbon dioxide from the organs and tissues back to the lungs.

Heritability.
The proportion of variation observed in a trait, such as height, that can be attributed to variation in inherited genetic factors.

Heterozygote (adjective: heterozygous).
An individual having two different alleles of a particular gene or genes.

Heterozygote advantage.
The case when there are two or more alternate alleles in a gene, and the heterozygote has a higher relative fitness than either homozygote.

Homozygote (adjective: homozygous).
An individual having two copies of the same allele of a particular gene or genes.

Human.
The term human refers to our species, *Homo sapiens*. It is sometimes also used in reference to other extinct species in our lineage, such as the Neanderthals and Denisovans, with the term *modern* humans reserved for our species. To avoid confusion, the term *hominin* (belonging to the family Hominidae and the subfamily Homininae) is a better description of our ancestral species.

Hygiene hypothesis.
The theory that abundant exposure to infectious and parasitic organisms during childhood may have played an important role in moderating our immune system, while today's more "sterile" environments set the stage for over-reaction of the immune system when it does encounter environmental or biological triggers.

Infertility.
The failure to achieve a clinical pregnancy after

12 months or more of regular unprotected sexual intercourse.

Innate immune system.
The body's first line of rapid defense against germs entering the body. It employs several mechanisms that are nonspecific, in that they respond in the same way to all germs and foreign substances.

Insertions/deletions.
A common type of mutation where one or more nucleotides are added to, or subtracted from a DNA base sequence, respectively.

Intelligence Quotient (IQ).
A score derived from a specific set of standardized tests developed to measure a person's cognitive abilities and indicate how far above or below they stand in comparison with their peer group.

Intracytoplasmic sperm injection (ICSI).
A form of in-vitro fertilization that involves injection of a single live sperm directly into an egg.

In-vitro fertilization (IVF).
A procedure in which eggs are removed from a woman's ovary and combined with sperm outside the body to form embryos. The embryos are grown in the laboratory for several days and then either placed in a woman's uterus or frozen for possible future use.

Lactose intolerance.
The inability to fully digest the sugar (lactose) in milk, usually caused by a deficiency of an enzyme called lactase. When lactose is allowed to proceed through the gut undigested, the bacteria in the colon exploit it, resulting in symptoms of abdominal discomfort, gas, bloating and diarrhea.

Macrophage.
A type of white blood cell that surrounds and kills microorganisms, removes dead cells, and stimulates the action of other immune system cells.

Major histocompatibility complex (MHC).
A group of immune genes located on chromosome 6 that have an important role in immune response. They code for proteins involved in the recognition of antigens, leading to destruction of infected cells, activation of macrophages and production of antibodies.

Mendelian diseases.
Diseases inherited in a simple pattern consistent with a mutation in a single gene.

Microbiome.
The community of microorganisms found living together in any given habitat. In humans, it refers to the abundant collection of mostly benign or beneficial microorganisms that colonize internal and external surfaces such as the skin, oral and vaginal mucosa, and the gastrointestinal tract – particularly the colon. **Microbiota** refers to the organisms composing the microbiome.

Mitochondria.
Numerous structures found within most cells that take in nutrients and break them down, producing energy that may then be used by the cell for a variety of functions.

Mutation.
A change in the DNA sequence of a cell, caused by mistakes during cell division or by exposure to DNA-damaging agents in the environment, such as radiation. Mutations can be harmful, beneficial, or have no effect.

Natural selection.

The process by which a heritable trait becomes either more or less common in a population based on the relative survival and reproductive success of those with the trait compared to others of the same species.

Positive selection.
Those individuals who happen to be best suited to an environment survive and reproduce most successfully, producing many similarly well-adapted descendants. Over time, the better-adapted come to dominate.

Negative, or purifying selection.
Those individuals with mutations that impair survival or reproduction will be less able to contribute to subsequent generations, so they tend to be eliminated from the population over time. New, harmful mutations will fail to become established.

Balancing selection.
Alternate variants of a gene can be perpetuated when the heterozygote is more fit than either homozygote, or when the fitness value of a variant increases when its frequency decreases.

Niche.

An ecological niche is the unique role and position a species has in its environment; how it meets its needs for food and shelter, survives, reproduces, and avoids competition with other species. **Niche construction** is the process by which an organism alters its own local environment to suit its needs, such as the building of nests and burrows by some animals. Humans have excelled at niche construction due to the combination of greater intelligence, manual dexterity, and cultural evolution.

Nucleotide.

A molecule consisting of a nitrogen-containing base (adenine, guanine, cytosine, or thymine (in DNA)/uracil (in RNA), a phosphate group, and a sugar (deoxyribose in DNA; ribose in RNA). DNA and RNA are composed of many nucleotides, strung together like beads in a necklace.

Pastoralism.
The practice of keeping sheep, cattle, or other grazing animals as the primary economic activity.

Phenotype.
All the observable characteristics of an organism resulting from the interaction of its genes with the environment; these include the organism's morphology, its developmental processes, its biochemical and physiological properties, its behavior, and the products of behavior.

Pleiotropy.
The situation where a single gene affects two or more apparently unrelated phenotypic traits.

Polygenic trait.
A characteristic, such as height or skin color, that is influenced by two or more genes.

Polygenic adaptation.
When polygenic traits are subject to positive selection a population may adapt through changes in allele frequencies at multiple genes, sometimes numbering in the hundreds or thousands.

Polygenic score.
A score that quantifies the sum of the effects of multiple genetic variants carried by an individual on a particular phenotypic trait such as height or intelligence. If the trait in question is a disease, a **polygenic risk score**

estimates the amount of overall increased risk of disease in an individual based on the combined effect of their particular spectrum of variants.

Protein.
One of at least 10,000 complex and often very large molecules, made up of long chains of hundreds to thousands of smaller molecules called amino acids, that serve an enormous number of essential roles in the human body.

Random genetic drift.
The change in frequency of an existing gene variant in a population over generations due to random chance. Random drift is more likely to occur in small populations.

Recessive.
Refers to the situation where a gene has two variants, or alleles. If the effect of one allele is masked by the other (the dominant one) when both are present, that allele is said to be recessive.

Relaxed selection.
The reduction or elimination of a source of natural selection that was formerly important for the maintenance of one or more traits.

Ribosomes.
Structures within cells where proteins are synthesized. Ribosomes "translate" the genetic code as they add amino acids to the developing protein molecule.

RNA, Ribonucleic acid.
A nucleic acid that has structural similarities to DNA; however, RNA is most often single-stranded and has a backbone made of alternating phosphate groups and the sugar ribose, rather than the deoxyribose. RNA contains information that has been copied from DNA and can

act as a messenger carrying instructions from DNA for directing the synthesis of proteins. There are also other forms of RNA, and each has a specific job in the cell.

Sexual antagonism.
Situation where the same genetic variant in different sexes may be subject to a different magnitude or even direction of natural selection.

Single nucleotide variant.
A DNA base sequence variation that occurs when a single nucleotide (adenine, thymine, cytosine, or guanine) is altered. For example, the base sequence AAG is a single nucleotide variant of AAT.

Structural variant.
A mutation where larger segments of DNA of varying lengths are altered, rather than just a single nucleotide. For example, segments of DNA may be inserted, deleted, copied, or inverted.

[1] The term *human* can apply to any species in the family Hominidae. A related term, *hominin*, usually refers to any species in the genus *Homo*.

[2] *Fitness* is a function of survival to reproductive age and reproductive success. I will discuss fitness and natural selection further in Chapter 4.

[3] *Heritability* is a genetic term concerning the nature of variation among individuals in a species. Considerations of 'nature vs. nurture' seek to identify those features of an individual that are determined intrinsically by its genetics (nature) versus those that are determined or modified by the environment within which the individual develops (nurture). Heritability refers to the former - that portion of

the observed variation which is due to underlying genetic variation. It is often estimated by establishing the degree of similarity between related individuals such as parents and their offspring, siblings, or twins.

[4] A chromosome is a thread-like structure found inside the nucleus of a cell. It is made up of proteins and DNA organized into genes.

[5] Watson and Crick were credited with this discovery and received the Nobel Prize in 1962. However, it is now generally acknowledged that another colleague named Rosalind Franklin also contributed significantly and did not receive the recognition she deserved at the time. She died of ovarian cancer before the Nobel Prize was awarded, and it is not given posthumously.

[6] For this reason, a mutation that changes ATT to ATC will be neutral, having no effect on the protein structure. We will discuss mutations in more detail later.

[7] individual who is heterozygous

[8] Haig D. 2003. The science that dare not speak its name. Quarterly Review of Biology 78(3): 327–35.

[9] There is another interesting angle to this story. The original allele at the regulatory gene site affecting TRPM8 has been found to have a strong protective effect against migraine headaches. The prevalence of migraine varies among different populations and is greatest in individuals of European descent – the population with highest frequency of the gene variant associated with cold adaptation. This may represent an instance of *antagonistic pleiotropy*, where positive selection for one trait (cold adaptation) has an adverse effect on another trait (protection from migraine). We will come back to this phenomenon in a later chapter when we address the question of whether our genome is perfect.

[10] A genome-wide association study, abbreviated GWAS. I will discuss this further in Chapter 6.

[11] The Greek root *homo-* means "same," and a *zygote* is a cell created when a sperm and an egg come together. The adjective for this situation is *homozygous*.

[12] Although the positive selection for lactase persistence is often assumed to have occurred around the same time that animal

domestication was introduced, one recent study cites evidence that the increased prevalence of lactase persistence may have occurred much more recently. (Burger, J. et al. 2020. Low prevalence of lactase persistence in Bronze Age Europe indicates ongoing strong selection over the last 3,000 years. Current Biology 30: 4307-4315)

[13] In reality, the genetics of hair color is much more complicated than this, involving multiple genes.

[14] Of course, the Holocaust associated with the second World War created a significant fourth bottleneck, although the genetic consequences cited in this paragraph were already established by that time. In fact, a genetic analysis of six recently excavated victims of antisemitic violence in 1190 revealed that Ashkenazi-associated disease variants were already present then.

[15] It is not that unusual for contemporary populations to be derived from a relatively small subset of ancestors, and it is interesting to consider why this is the case. It is likely that some of the initial colonizers were already beyond reproductive age. Among the others, differential success in reproduction and choice in the number of children are also very likely. Even relatively small differences in fertility can amplify over time, especially if tendencies toward more children are passed along (culturally or genetically) over generations.

[16] A single nucleotide variant refers to the case where one single DNA base is substituted for another, e.g., the sequence AAG is a single nucleotide variant of AAT.

[17] Fu, Wenqing et al. 2013. Analysis of 6,515 exomes reveals the recent origin of most human protein-coding variants. Nature 493(7431): 216–220.

[18] Muller, H. J. 1950. Our load of mutations. The American Journal of Human Genetics 2(2): 111-176.

[19] Benton, Mary Lauren et al. 2021. The influence of evolutionary history on human health and disease. Nature Review Genetics 22: 269–283.

[20] Nurk, Sergey et al. 2022. The complete sequence of a human genome. Science 376(6588): 44-53.

[21] This was a 7-year international research effort launched in January 2008, with the goal of establishing the most detailed catalog

of human genetic variation to date. It was designed to detect variants with frequencies as low as 1%. DNA was analyzed from 2504 individuals from 26 populations drawn from five continental regions - East and South Asia, Europe, Africa, and the Americas. The 1000 Genomes Project Consortium 2015. A global reference for human genetic variation. Nature 526: 68–74.

[22] Halldorsson, Bjarni V. et al. 2022. The sequences of 150,119 genomes in the UK Biobank. Nature 607: 732-740.

[23] *Structural variants* is a general term that applies to a mutation where larger segments of DNA of varying lengths are substituted, rather than just a single nucleotide. It is somewhat subjective; variously defined as involving a minimum of 50 base pairs vs. a minimum of 1,000 base pairs. Although the least common, structural variants affect the greatest number of bases due to their large size. They have been estimated to account for 15% or more of human disease-causing mutations. *Insertions* and *deletions* refer to segments of DNA that are added to, or subtracted from a DNA sequence, respectively.

[24] E.g., recall how the gene variant for tusklessness in elephants improved survival for females, but resulted in early mortality for males.

[25] Claussnitzer, Melina et al. 2020. A brief history of human disease genetics. Nature 577: 179-189.

[26] https://en.wikipedia.org/wiki/Disease. Accessed on 1/16/2022.

[27] Stenson et al. 2020. The Human Gene Mutation Database (HGMD®): optimizing its use in a clinical diagnostic or research setting. Human Genetics 139(10): 1197-1207. http://www.hgmd.cf.ac.uk/ac/index.php. Accessed January 10, 2023.

[28] Ruderfer, Douglas M. et al. 2016. Patterns of genic intolerance of rare copy number variation in 59,898 human exomes. Nature Genetics 48(10): 1107-1111.

[29] This is an unusually large effect for a single variant in a typical GWAS.

[30] We encountered this term previously with regard to *polygenic adaptation*.

[31] Dobzhansky, Theodosius 1961. Man and natural selection. American Scientist 49(3): 285-299.

[32] Rendell, Luke et al. 2011. Runaway cultural niche construction. Philosophical Transactions of the Royal Society B 366: 823–835.

[33] Jeffery, W.R. 2005. Adaptive evolution of eye degeneration in the Mexican Blind Cavefish. Journal of Heredity 96(3):185-196.

[34] M. E. McPhee & N. F. McPhee 2012. Relaxed selection and environmental change decrease reintroduction success in simulated populations. Animal Conservation 15:274–282.

[35] Rühli, Frank Jakobus and Maciej Henneberg 2013. New perspectives on evolutionary medicine: the relevance of microevolution for human health and disease. BioMed Central Medicine 11:115. http://www.biomedcentral.com/1741-7015/11/115

[36] World Health Organization 1972. Genetic disorders: prevention, treatment, and rehabilitation. Technical Report Series No. 497. https://apps.who.int/iris/bitstream/handle/10665/40966/WHO_TRS_497.pdf?sequence=1&isAllowed=y

[37] Michael Lynch 2016. Mutation and human exceptionalism: Our future genetic load. Genetics 202: 869–875.

[38] Chew, Nicholas W.S. et al. 2023. The global burden of metabolic disease: Data from 2000 to 2019. Cell Metabolism 35: 414–428.

[39] These include cancers in the breast, colorectum, endometrium, esophagus, extrahepatic bile duct, gallbladder, head and neck, kidney, liver, bone marrow, pancreas, prostate, stomach, and thyroid. Ugai, Tomotaka et al. 2022. Is early-onset cancer an emerging global epidemic? Current evidence and future implications. Nature Reviews Clinical Oncology 19(10): 656-673.

[40] Staub, Kaspar et al. 2018. Increasing variability of body mass and health correlates in Swiss conscripts, a possible role of relaxed natural selection? Evolution, Medicine, and Public Health 2018: 116–126.

[41] Body mass index (BMI) is a measure of body fatness, calculated as a person's weight divided by the square of their height.

[42] You, Wenpeng and Maciej Henneberg 2018. Relaxed natural selection contributes to global obesity increase more in males than in females due to more environmental modifications in female body mass. PLoS ONE 13(7): e0199594. https://doi.org 10.1371/

journal.pone.0199594.

[43] You, Wenpeng and Maciej Henneberg 2016. Type 1 diabetes prevalence increasing globally and regionally: the role of natural selection and life expectancy at birth. British Medical Journal Open Diabetes Research and Care 4: e000161. https://drc.bmj.com/content/bmjdrc/4/1/e000161.full.pdf.

You, Wenpeng and Maciej Henneberg 2018. Cancer incidence increasing globally: The role of relaxed natural selection. Evolutionary Applications 11: 140–152.

You, Wenpeng et al. 2022. Healthcare services relaxing natural selection may contribute to increase of dementia incidence. Nature Scientific Reports 12: 8873. https://doi.org/10.1038/s41598-022-12678-4.

[44] Fuchs, Ivan 2019. The Evolutionary Mechanism of Human Dysfunctional Behavior. Radius Book Group: New York.

[45] Michael Lynch 2016. Mutation and human exceptionalism: Our future genetic load. Genetics 202: 869–875; James F. Crow 1997. The high spontaneous mutation rate: Is it a health risk? Proceedings of the National Academy of Sciences USA 94: 8380–8386; James F. Crow 1968. Rates of genetic change under selection. Proceedings of the National Academy of Sciences USA 59: 655-661; Adam Eyre-Walker, Megan Woolfit and Ted Phelps 2006. The distribution of fitness effects of new deleterious amino acid mutations in humans. Genetics 173: 891–900; respectively.

[46] Muller, H. J. 1950. Our load of mutations. The American Journal of Human Genetics 2(2): 111-176.

[47] McQueen, Hannah 2022. The 10 Most Expensive Drugs in the US, Period. https://www.goodrx.com/healthcare-access/drug-cost-and-savings/most-expensive-drugs-period. Accessed June 13, 2022.

[48] Genetic variance has been estimated to account for 28% of the variation in age at first sexual intercourse, 10-55% of variation in age at first reproduction, 15% of variation in the total number of children ever born, 45% of variation in the age at menopause, and 24-39% of the variability in overall reproductive success. Age at menarche has received the most attention, with an estimated heritability in the range of 50-70%, likely associated with hundreds of genetic variants.

[49] Ronald Lee 2003. The demographic transition: Three centuries of fundamental change. Journal of Economic Perspectives 17(4): 167–

190.

[50] From Sohn, Emily 2020. Planning for success. Nature 588: S162-S164.

[51] We'll go into more detail about the evolution of intelligence in Chapter 12.

[52] Infertility is defined by the failure to achieve a clinical pregnancy after 12 months or more of regular unprotected sexual intercourse.

[53] CDC 2023. https://www.cdc.gov/reproductivehealth/infertility/index.htm. Accessed March 19, 2023.

[54] Jiang, Ziru et al. 2017. Genetic and epigenetic risks of assisted reproduction. Best Practice & Research: Clinical Obstetrics and Gynaecology 44: 90-104.

[55] Oates, Robert D. et al. 2002. Clinical characterization of 42 oligospermic or azoospermic men with microdeletion of the AZFc region of the Y chromosome, and of 18 children conceived via ICSI. Human Reproduction 17(11): 2813–2824.

[56] Czeizel, Andrew E. and Kenneth J. Rothman 2002. Does relaxed reproductive selection explain the decline in male reproductive health? A New Hypothesis. Epidemiology 13(1): 113-114.

[57] Hanevik, Hans Ivar et al. 2016. Can IVF influence human evolution? Human Reproduction 31(7): 1397–1402.

[58] Black Death was caused by the plague bacterium *Yersinia pestis* that was present in the populations of fleas that infested rodents.

[59] McManus, Kimberly F. et al. 2017. Population genetic analysis of the DARC locus (Duffy) reveals adaptation from standing variation associated with malaria resistance in humans. PLoS Genetics 13(3): e1006560. https://doi.org/10.1371/journal.pgen.1006560.

[60] Klunk, Jennifer et al. 2022. Evolution of immune genes is associated with the Black Death. Nature 611: 312-319.

[61] This same favored variant is known to be a risk factor for Crohn's disease and other infectious diseases. One of the other four variants is also currently associated with the risk of rheumatoid arthritis and systemic lupus erythematosus. This chapter goes into more detail on evolutionary tradeoffs and autoimmune diseases.

[62] This statement is true, but the situation is not static. Infections caused by antibiotic-resistant bacteria are increasing, especially in lower-income countries. In 2019, antibiotic resistance was estimated

to be directly responsible for 1.27 million deaths globally, exceeding the number from HIV/AIDS or malaria [Antimicrobial Resistance Collaborators 2022. Global burden of bacterial antimicrobial resistance in 2019: a systematic analysis. The Lancet 399: 629-655]. The malarial parasite in Africa was recently found to be resistant to a major family of drugs used to treat infection [Kozlov, Max 2021. Resistance to key malaria drugs confirmed in Africa. Nature 597: 604].

[63] Crow, James F. 1968. Rates of genetic change under selection. Proceedings of the National Academy of Sciences 59: 655-661.

[64] Stiemsma, Leah T. et al., 2015. The hygiene hypothesis: current perspectives and future therapies. ImmunoTargets and Therapy 4: 143–157.

[65] Rook Graham A. W. 2012. Hygiene hypothesis and autoimmune diseases. Clinical Reviews in Allergy & Immunology 42: 5–15.

[66] Microbiota refers to all the micro-organisms in our environment.

[67] Vatanen, Tommi et al. 2016. Variation in microbiome LPS immunogenicity contributes to autoimmunity in humans. Cell 165: 842–853.

[68] Our World in Data, https://ourworldindata.org/explorers/coronavirus-data-explorer. Accessed February 18, 2023.

[69] Souilmi, Yassine et al. 2021. An ancient viral epidemic involving host coronavirus interacting genes more than 20,000 years ago in East Asia. Current Biology 31: 1–11.

[70] COVID-19 Host Genetics Initiative 2021. Mapping the human genetic architecture of COVID-19. Nature 600: 472-477.

[71] Kousathanas, Athanasios et.al. 2022. Whole genome sequencing reveals host factors underlying critical Covid-19. Nature 607: 97–103.

[72] Zeberg, Hugo and Svante Pääbo 2020. The major genetic risk factor for severe COVID-19 is inherited from Neanderthals. Nature 587: 610-612.

[73] For an interesting article that highlights some of the controversies in this field, see the September 13, 2021 issue of The New Yorker, "Can progressives be convinced that genetics matters?" by Gideon Lewis-Kraus.

[74] Down syndrome results from the inheritance of three copies of chromosome 21, rather than the usual two copies.

[75] Lee, James J. et al. 2018. Gene discovery and polygenic prediction from a genome-wide association study of educational attainment in 1.1 million individuals. Nature Genetics 50: 1112-1121.

[76] Polygenic scores are derived from genome-wide association studies that identify correlations between specific genetic variants and traits such as intelligence. The calculation of the scores is based on the sum of the effects of all such variants carried by a particular person. Polygenic scores have been used to predict phenotypes such as intelligence or risk of disease based on the composition of a person's genome. However, the utility of these scores is not universally accepted. We will discuss them further in the next chapter.

[77] Belsky, Daniel W. et al. 2016. The genetics of success: How single-nucleotide polymorphisms associated with educational attainment relate to life-course development. Psychological Science 27(7): 957–972.

[78] Peach, Hannah et al. 2014. Replication of the Jensen effect on dysgenic fertility: An analysis using a large sample of American youth. Personality and Individual Differences 71:56–59.

[79] Meisenberg, Gerhard 2010. The reproduction of intelligence. Intelligence 38:220–230.

[80] Woodley of Menie, Michael A. 2015. How fragile is our intellect? Estimating losses in general intelligence due to both selection and mutation accumulation. Personality and Individual Differences 75:80–84.

[81] Kong, Augustine et al. 2017. Selection against variants in the genome associated with educational attainment. Proceedings of the National Academy of Sciences 114(5): E727-E732. https://doi.org/10.1073/pnas.1612113114

[82] Jalovaara, Marika et al. 2018. Education, gender, and cohort fertility in the Nordic countries. European Journal of Population 35: 563–586.

[83] Hazan, Moshe and Hosny Zoabi 2014. Do highly educated women choose smaller families? The Economic Journal 125: 1191–1226.

[84] Pontzer, Herman 2023. The human engine. Scientific American 328(1): 24-29.

[85] The average annual cost for treatment of hemophilia B using factor IX administration is $700,000-800,000.

[86] The cells at this stage are called "chimeric antigen receptor T-cells," abbreviated CAR-T. This abbreviation is also used when referring to this type of therapy.

[87] The doctor who led the team has been harshly criticized by the medical community for violating experimental and ethical standards; he was fired from his position and was recently released from prison.

[88] For example, see: https://www.technologyreview.com/2019/03/08/136730/23andme-thinks-polygenic-risk-scores-are-ready-for-the-masses-but-experts-arent-so-sure/.

[89] Turley, Patrick et al. 2021. Problems with using polygenic scores to select embryos. New England Journal of Medicine 385(1): 78-86.

[90] Susan M Holloway and Charles Smith 1975. Effects of various medical and social practices on the frequency of genetic disorders. Am J Hum Genet 27: 614-627.

[91] Mossburg, Cheri 2023. CNN article published online March 2, 2023. https://www.cnn.com/2023/03/02/us/california-embryo-cancer-gene-lawsuit/index.html.

[92] Nathaniel Comfort 2012. The eugenic impulse. The Chronicle Review, November 12, 2012.

[93] Eric S. Lander 2015. Brave new genome. N Engl J Med 373(1): 5-8.

ACKNOWLEDGEMENTS

I am indebted to my wife, Patricia Blau, for years of patience allowing me to spend much time on the development of this book, and for her willingness to listen to my triumphs and frustrations along the way. Patricia Blau, Sarah Blau, Jim Martin, and Bruce Boehm spent some of their valuable time reading portions of this book and providing critical feedback. Thanks also to my editor Haley Paskalides, and cover designer Jess Estrella.

ABOUT THE AUTHOR

William S. Blau Md, Phd

William S. Blau is an Emeritus Professor at the University of North Carolina School of Medicine at Chapel Hill. He was raised in New Hyde Park, New York. After obtaining a Bachelor of Science degree at the State University of New York at Stony Brook, he enrolled as a graduate student at Cornell University in Ecology and Evolutionary Biology. Under the mentorship of Dr. Paul Feeny, he spent a year conducting field research in Costa Rica and went on to complete his Doctorate in 1978. Two subsequent postdoctoral positions brought him to Chapel Hill, North Carolina, where he enrolled in Medical School in 1983. He remained at the University of North Carolina for a residency in Anesthesiology, specializing in Pain Medicine. He joined the faculty at UNC Chapel Hill in 1991, serving as the Division Chief for Pain Medicine and establishing an accredited pain management fellowship. He retired from UNC in 2015 and remains in Chapel

Hill with his wife, Patricia. When not writing, he enjoys traveling, visiting his two adult children and granddaughter, volunteering at UNC Hospitals and with the Triangle Land Conservancy, and performing as one half of the musical duo Blumora.

Contact the author at: bblau@email.unc.edu

Printed in Great Britain
by Amazon